Fisheries Extension

ABOUT THE BOOK

Fisheries not only gives nutritional security to people, it also provides livelihood, to millions of people the extension communication/media are the great part of any work/research/study without which no body knows what is going on in this changing world. The book deals with fishery extension, communication, communication process, training, training management project, project formulations. Project preparations shrimp, carp cum prawn farming, its feasibility sensetivity, entrepreunership, ecosystem structure and some models. A comprehensive knowledge of fishery extension, training and enterpreneurship has been given in this book to link farmers, students, trainers, extension workers, teachers and entrepreneurs to achieve the goal of maximum production and employment.

Fisheries Extension

Author
Amita Saxena
Professor, College of fisheries
GBPUA&T
Pantnagar – 263 145

2011
DAYA PUBLISHING HOUSE®
New Delhi - 110 002

Published by : **Daya Publishing House**
A Division of
Astral International Pvt. Ltd.
– ISO 9001:2008 Certified Company –
4760-61/23, Ansari Road, Darya Ganj
New Delhi-110 002
Ph. 011-43549197, 23278134
E-mail: info@astralint.com
Website: www.astralint.com

Laser Typesetting : **Twinkle Graphics**
Delhi

Printed at : **Chawla Offset Printers**
Delhi - 110 052

PRINTED IN INDIA

CONTENTS

Preface .... *v*

1. Fisheries Extension 1-25

2. Communications 26-30

3. Communication Process 31-51

4. Training 52-65

5. Management of Training Programme 66-72

6. Project 73-78

7. Project Formulation 79-91

8. Organic Shrimp Farming and Sustainability 92-95

9. Preparation of Project on Establishment of Shrimp Hatchery 96-106

10. Shrimp Farming in India 107-110

11. Preparation of Project on Carp-Cum-Prawn Culture 111-113

12. Indian Aquaculture and Evaluation of Investment Feasibility 114-130

13. Sensitivity Analysis 131-135

14. The Entrepreneurship 136-154

15. Entrepreneurial Plan for Aquarium Business 155-202

16. Ecosystem Structure 203-211

17. Some Models Individual Based 212-228

PREFACE

Fishery sector is an important source of National revenue. The aim of Fishery Extension to extend lab to land programme, to introduce advance scientific research/knowledge for the benefit of farmers/fisherman by training, technical consultation, audiovisual aids, workshops, visit to the spots (farmers'/fisherman place), demonstrations, exhibitions etc. In this way, the fishery production will reach to maximum. It promotes the over all development of fishery sector at National level.

For this funds come from various levels including Centre and State financial subsidies, development programme funds, non Government Organisation International funds/loans, MPEDA, Private funds Enterprises etc.

This book covers all above aspects under the chapter title Training and Management, Policies and Acts of Fisheries Extension Education, Communication, Projects and project formulation, Shrimp, Carp projects, Project Evaluation, Sensetivity Analysis, Entrepreneurship Ecosystem Structure, Models etc.

This book is useful for students, planners, policy makers, company executives, research scientists workers and politician also.

Author is thankful to elders, senior, professors, scientists DGICAR, Dean Hon'ble Vice Chancellor, students. Library for providing necessary help and literature/information to write this book. My due thanks to Daya publishers who published this work without whom no one can see this piece of work.

Amita Saxena

VOL - I
FISHERIES EXTENSION

Chapter 1

FISHERIES EXTENSION

Project

A project is a scheme to organize the use of a given quantity of resources in a specific way to achieve particular results within a definite time period. The execution of a project requires multidisciplinary efforts mobilising different skills and resources to achieve predetermined development objective directly or indirectly, in new or added value or social economic and financial organisation benefits.

Within this general definition many different examples of projects can described each different type of organisation however they all have the same fundamental pages for the time each project is Ist conceptualize until it is terminated, these parts fit together and relate to each other which is called as project cycle.

The Relationship between Project and Development Plans

Project may arise from many resources they may originate or we included within this sector study and from their the translated into the sector, development plans.

Many of the weakness in projects which materialized in project implementation may be attributed to poor project formulation, in particular the often less than systematic way in which project ideas are integrated into the economic and institutional fabric of the sector thus good project formulation starts at the sectural planning stage and the keep to good sectural planning is sector study.

The sector study consists of two parts :

1. Sector Review of stock taking
2. Sector Analysis a diagnosis

Strategy

The strategy is key planning document. It may be appropriate for a country aquaculture section to have a signal strategy, ultimately if there are no section in that area (such as shrimps and sea weeds) or there are different climatic zones (such as upland areas, coastal areas) then each may require different approaches and development initiating of a separate strategy may be needed for each sub-sector.

Major Research Project and Programe for the Near Future

Capture Fisheries (Inshore and Offshore fisheries).

1. The delineation of some sps/stocks/population in all the sps. supporting the major fishery and determination of the rates of their intermingling in space and time.

2. Re-examination of the stock assessment estimates the important species/groups made so far in the life of the presence or intermingling of the different other communities of the fish with each sps. in space and time.

3. Continuation of the stock assessment year by year for imparting advice on exploitation strategy.

4. Calculation of the fecundities of a different important species in respect to space and time and estimation of overall egg production potential of a particular species or sub species with respect to space and time.

5. Monitoring the feeding intensity of the major species in relation to major food items present in the environment and evaluation of relationship between the abundancy and availability of the fish with respect to major food items.

6. Determination and monitoring of the relationship between the parental stock and the progenies of the important species in space and time.

7. Spawning survey and egg census of the major species and groups in space and time.

8. Development of statical model between the egg potential of different species and the strength of zero year class to other older age group.

9. Correlation of important oceanic parameter in the surface mid water and bottom water—area of fishing ground with respect to availability and abundance of major species and resources in space and time.

10. Development of statical model on the relationship between various oceanographic parameters and the availability and abundance of the various pelagic or midwater or dimersal species in space and time.

11. Integration testing and refining of the models data developed from the above projects for each and every commercially important species for effective forcasting and their annual or seasonal fishery.

12. Acoustic survey and experimental fishing in the deep sea area to assess the availability of fish stock there.

13. Preparation of regionwise, areawise and seasonwise charts and maps for optimum exploitation of all the species.

Fishery Oceanography and Environmental Research

1. Collection and Analysis of uninterrupted data and monitoring of the various physical, chemical and biological parameters in the fishing grounds at surface, midwater and bottom levels.

2. Experimental study on the impact of the bottom travelling on the quality and quantity of the benthic biota in the fishing grounds.

3. Evaluation of the impact of various mesh sizes of different gears on various segment of the population such as juvenile and boarders.

4. Identification of early developmental stages of the all major species and groups supporting the fisheries for egg census and spawning survey.

5. Assessment of mesopelagic resources in EEZ.

6. Collection and consolidation of all the data of ecological parameters of coastal water bodies estuaries, and lagoons system, as well as bringing out maps, charts and status ratio on the extent of the threatened/endangered, vulnerable and rare species as well as threatened environment.

7. Monitoring of various forms of industrial or agriculture or any other kind of pollution for suggesting remedies.

8. Study on the various flora and fauna of the coastal zone for evaluating biodiversity and conservation.

9. Base line study with regards to project and programme that need environmental impact assessment.

Imparting advice/consultancy or assessment to the agency such as Govt. department industrial concerns, entrepreneurs, planners and administrator on the various aspects of judicial coastal zone management.

Stages of Project Cycle

There are mainly 5 stages in the cycle of existence of the project these are :

1. Identification
2. Project Preparation
3. Appraisal and Agreement
4. Implementation and monitoring
5. Evaluation
6. Diagram of Project Cycle

1. Identification : Project idea is translated into a preliminary primary description of project.

(*a*) Extent and limits of project or proposed

(*b*) Different approaches to project are identified

(*c*) Judgement made regarding which option should be taken forward to project preparation.

2. Project Preparation

(*a*) At this stage project is designed.

(*b*) Objectives input output organisation, participants are defined.

(*c*) Cast and earning are calculated

(*d*) Financial Forum are prepared

(*e*) Expected results is analyzed

(*f*) Social economic and environmental impacts are estimated

(*g*) Provisional and final project documents are prepared

3. Appraisal and Agreement

(*a*) Project are prepared from previous project documents.

(*b*) Success of meeting, clearance and financing negotiation take place

4. Project Implementation

(*a*) Project management and lines of command are established

(*b*) Various implementation procedure established.

(*c*) In course of implementation, project progress is monitor and revised and adoptions are made for unexpected events and finally project is gone to go competition.

5. Project Evaluation

Project evaluation takes place at a stable time after the project has been completed, project objectives. Project implementation and project benefits are revised. This evaluation may result in the project been exchanges or in the identification of new project and they need to release of method by which similar project will be formulated to the future.

An Overview of Project Formulation

The first three stages of project cycle are undertaking in sequence but work can be broken into many different ways the constituent parts may be called phases which are further broken into stages, phases and steps can be overlapping some degree each step contains one or more task.

Stages	Phases	Steps
1. Identification	1. Preparation for project formulation	1. Project inspection
	2. Reconnaissance and primary project design	2. Preparation of formulation work plan.
		1. Analysis of situation from an overall perspective.
		2. Analysis of the situation having regard to main interest group involved.
		3. Assessing the future without project
		4. Outline specification of all possible project.
2. Preparation	1. Project design	1. Detail technical and socioeconomic investigation
		2. Definition of projects objectives, targets and design criteria.
		3. Design of individual project component
		4. Design of project organisation structure management arrangement.
		5. Project cost revenues and financing proposal
	2. Analysis of expected results	1. Financial Analysis
		2. Economic Analysis
		3. Social Analysis
		4. Environmental Impact Analysis
	3. Project submission	1. Project documentation and submission.
3. Project Appraisal	1. Project Negotiation	2. Project appraisal and negotiation.

Ist Phase

Preparation for project formulation :

This phase concern all the activities necessary to prepare for formulation of project. It has two steps :

1. Project inception.
2. Preparation for formulation plan.

The output is a programme of work for project formulation.

IInd Phase

Reconnaissance and Primary Project Design.

IInd phase contains all activities necessary to define objective of project identification and consideration for meeting these objective making a primary assessment of content of project and its likely affects. It normally has four steps :

1. Analysis of situation from an overall perspective.
2. Analysis of situation from perspective of main interest group involve.
3. Assessing the future without project
4. Outline specification of possible projects.

The output of 2nd phase is primary design of project including justification of its main features such as location, type of participants, main activities size, timing organisation structure and management system is also include Ist estimation of cost at this point a project report is normally submitted to funding organisation for approval before further formulation work is undertaken.

IIIrd Phase : Project Design

This phase normally initiates formal project.

Preparation typically follows five steps :

1. Detail technical and socio-economic investigation.
2. More precise definition of project objectives target and design criteria.
3. Design of individual project component.
4. Design of project organisation structure and arrangement.
5. Project cost and revenues estimation its financing proposal.

The output of phases in full description and casting of project together with financing plan.

IVth Phase

Analysis of expected results.

This phase concern all activities necessary to assess the project. The work typically contains 4 stages.

1. Financial Analysis
2. Economic Analysis
3. Social Analysis
4. Environment Impact Analysis

The output of this phase is determination of effect and impact of project.

Vth Phase

Project documentation and submission :

This phase concern or activities necessary to repair a final project document complete with design and relevant analysis the function of 5th phase may be described in one step. Project documentation and submission. The output of the phase is projected document.

VIth Phase : Negotiating the Project

This phase concern all activities necessary to have the project document accepted and the project finance for implementation. It starts when source of financing accept the project formulation documented the work include only one step. Project appraised and negotiation the output is projectfully ready for implementation under proper administration and with necessary financial commitments.

Different Task in Steps of Project Formulation

Each step has different task.

1. Project Inception

Task 1st

1. Idea development.
2. Preparation of logical framework.
3. Recruitment and mobilization of formulation team.
4. Receive of assignment and review of formulation team.
5. Provisional project concept.

2. Preparation of Formulation Work Plan

1. Project planning, meeting.
2. Prepare work plan and allocate responsibilities to the biologist, ecologist, engineers and sociologists.
3. Work plan for formulation.

3. Analysis for diagnosis of situation

1. National, Regional and securial background.
2. General, evaluation of project area.
3. Identification of possible course of action.

4. Analysis of project having regard to the people involved

1. Identification of interested group.
2. Detail study of agent.
3. Identification of possible course of action.

5. Step Assessing the Future without Project

1. Projecting demographic changes.
2. Projecting demand, supply and price.
3. Definition of without project situation.

6. Outline specification of possible project

1. Classification of a project option.
2. Analysis of alternative selection.
3. Preparation of Final Project document.

7. Detail Technical and socio-economic investigation

1. Necessary technical and socio-economic investigation of project.

8. Definition of Project, Objective, Target and Design

1. Review and Refinement of Objective.
2. Specification project target.
3. Confirmation of define.
4. Preparation of work plan.

9. Design of Individual Component

1. Design of production component.
2. Design of production support component.
3. Design of social support component.
4. Phasing of component and scheduling.

10. Project Organisation and Management

1. Deciding organisation structure.
2. Designing a system for project monitoring and evaluation.

11. Project Cost and Revenues Estimation and Financing Proposal

1. Completing estimation of individual component cost.
2. Estimating project revenues.
3. Preparing financing plan.

12. Financial Analysis

1. Analysis at farm level.
2. Analysis at level of credit institution.
3. Analysis at level of project entity.
4. Analysis of effect of Govt. budget.

13. Economic Analysis

1. Impact on economic growth.
2. Impact on foreign exchange.
3. Impact on economic distribution.

14. Social Analysis

1. Production organisation.
2. Population management settlement pattern.
3. Stand living indicators.

15. Environmental Impact Analysis

1. Alteration in natural features.
2. Conservation measures and management of revenable resources.

16. Preparation of the Project Report and Submission

1. The report structure.
2. Content of report.
3. Report submission and presentation.

17. Project Negotiation

1. Preliminary negotiation.
2. Project appraisal
3. Fulfillment of conditions by Government.
4. Project agreement.

Special Characteristics of Fisheries Project

In aquaculture production project need of water resources is the basis requirement. This requirement immediately limit options for :

(a) The species which can be formed as tolerant to fresh water, brackish water, Marine environment.

(b) The system which can be used as extensive, semi-intensive, intensive system and all depends upon availability of resources and input.

(c) The practices which can be used as all farm units such as pond, raceway, floating cages and raft etc. each have a characteristic which make them particularly applicable in certain condition. These all parameters influence the type of project which can be formulated, project with the activity which extent over large areas are new characteristics of aquaculture development. Projects are typically and relatively small and highly specific in terms of their objectives.

Types of Aquaculture Project

1. **Private Sector Project :** When investment is by commercial interest *e.g.,* a shrimp farming company wants to build a new hatchery.

2. **Public Sector Project :** Where the investment is in a publically owned and entity for *e.g.,* the ministry of fishery wants to build a state hatchery to support its programme to enhance inland fishery.

3. **Public Sector Project :** Where investment is by private farmers supported by Govt. services *e.g.,* ministry of agriculture wants to increase inland fishery production.

4. **Public Sector Project :** Public Sector Project concern with institutional building *e.g.,* the ministry of agriculture need to improve its organisation and management of aquaculture sub-sector.

Problems in Fisheries Project

Various problems of fisheries project can be grouped under five broad headings :

1. Institutional Difficulties

2. Socio-economic Aspects

3. Market and Marketing

4. Fish Resources

5. Appropriate Technology and Provision of Credit

1. Institutional Difficulties

(a) *National Policies*

(i) Lack of foreign exchange to purchase essential input.

(ii) Absence of legislations to manage control and project fisherman.

(iii) Poor coordination between national institutions.

(b) Inadequate technical input at identification and preparation stage

(c) *Inadequate use of technical assistance*

(d) *Delay in project implementation* : The staff of the implementing agency may not be familiar with procurement of the funding agency.

(e) Poor project management because of lack of suitable and trained staff and sometime because of complete between staff members of different two agencies.

2. Socio-Economic Aspect

Sometimes financing agencies have been criticized for putting emphasis on return and investment and paying little attention to social factors therefore a considerable amount of time may be required for collecting necessary sociological and economic data.

3. Market and Marketing

Some project have been failed because of an inadequate understanding of cost and prices.

4. Fish Resources

Catch forcasting is difficult it requires scientific data on environmental factors.

5. Appropriate Technology and Provision of Credit

The technical design of fisheries component has not always been satisfactory due to which project have been failed or delay lack off shore facilities also a problem of repayment of loan is also a problem in aquaculture industry.

Project Report Preparation

A project document gives a detail account of approach means and action plan for achieving the objective of project. The key ideas of project report are approach means and action plan.

Approach

It is to identify organised technological aspects of implementation of action plan of project.

Means

Identify the requirement of resources like land, water, man's power and other material.

Action Plan

Strategy to achieve the target. It gives the detail of different strategies in order of their priorities and represent the implementation and monitoring mechanism. It also includes care required from the legal point of view.

Project Classification

Project basically can be classified under two categories :

1. Social development project.
2. Investment oriented commercial project.

1. Social Development Project

SDP are funded by the national or international development agencies. They are directed mostly to upgrade the resources. *e.g.,* such project in fishery sector are :

   (*a*) World Bank project in Inland Fisheries Development for establishment of carp hatchery in 5 states of India.

   (*b*) Project for development and shrimp culture and lake fisheries.

2. Investment Oriented Commercial Projects (IOCP)

These projects are commercial in *nature earning a desirable rate of returns is the main objective of these type of project.*

These projects may be classified under two groups :

   (*a*) The small scale operated oriented project where the operator executes the project to earn livelihood and to earn on income by sailing fish or fishery product after making cost of production.

      If bank funded project and small fisherman, backyard hatcheries, village level fish production unit etc. The return of these projects are evaluated by cash income.

   (*b*) The investment oriented project for enterprises. In these projects entire cost including management and operation are explained by expenditure account. The returns are evaluated by definite financial framework.

For example, shrimp culture of different private companies of India and fish processing plant establish by private or semi Govt. agencies.

Elements of Project Document or Project Report

The essential elements are :

1. **Title :** What, where, why must be cleared to the title itself it has to be fairly communicated *e.g.,* establishment of 10 million mini shrimp hatchery for easy availability of shrimp seed in Kakinadu district of A.P.

2. **Background :** It gives a brief account of production demand available resources and market status.

3. **Objective :** Objective should be clear.

4. **Technology Selection :** They are various concerning points for selection of appropriate technology such as agroclimatic suitability timely supply of input, service and man power.

5. **Site Selection :** Site selection is to be done on the basis of concerning aspect as technological aspect, legal aspect and marketing aspect.

6. **Project Input :** Input of a project are classified under different gps which indicate qualities require including their source of availability *e.g.,* (*i*) capital input, (*ii*) operational input, (*iii*) fixed cost.

   (*i*) *Capital Input :* As land water body building equipment and other farm machinery.

   (*ii*) *Operational Input :* Supply of brooder feed, chemicals, packaging material.

   (*iii*) *Fixed Stock :* Technically skilled manpower.

7. **Output :** It indicates the product output in time and quality in different years.

8. **Cost :** This elements cover the cost of capital items, operational cost, marketing cost including advertisement, and cost of item with tax and subsidiary.

9. **Activity Time Frame :** A clear time frame of all activities is need *e.d.,* for implementation and monitoring.

10. **Source of Earn :** This includes the source of finance.

11. **Financial Evaluation**

The Report Structure

An excellent way for judging what should be tenth of different parts of report their score and detail require is to identified who is going to all the report or only a certain part of it. Typically divided sections are :

1. Summary, 2. Main Report, 3. Appendixes, 4. Working Plan.

1. **Summary :** The summary provides a self content picture of project and all its implications understandable without further reference and it is not more than 2-3 pages this may be given to the top policy maker such as to minister or chief executive.

2. **Main Report :** The main report is written in consistant style with well balance section, it is an executive summary of project based on materials presented in finding and conclusions.

   This might be for non-technical senior managers permanent secretaries in Govt. and all interested parties.

3. **Appendixes :** Each appendixes is a self content analysis of a single measure aspect of proposed project, these are usually read with considerable thoroughness by head of technical department and technical specialist.

4. **Working Plans/Papers :** Each working papers is essentially a fill on basic data estimates and calculations on which the alternative design for project component in the respective appendix are analysed.

Content of the Main Report

Content of the main report are organised in sections arrange in sequence and for the production project usually include the following :

(*a*) Summary

(*b*) Introduction

(*c*) Background

(*d*) The project area or sub-sectors

(e) Proposed Project

(f) Organisation and management

(g) Market and Prices

(h) Financial or Economical Implication

(i) Project Justification and risk

(j) Outstanding issues and follow up action require.

1. Summary

In this one should describe the brief account of his proposed with and what are his expectations.

2. Introduction

In main report the purpose of introductory section is to provide the contest and setting within which the investment proposal have been formulated.

3. Background

The background section introduces important features of current policy on aquacultural development reviewers recent trend in relevant sector and describe the organisation and function of major institution concern.

4. Project Area Subsector

Project area subsector prepares the ground for explaining the project design decision during formulation.

5. The Proposed Project

The proposed project including a coincise descriptions of all major features of project and presents a summary cost estimate and financing plan and explains recommend procurement accounting and auditing procedure.

6. Organisation and Management Section

This section explains how various institution and agency will participate in implementing the project and operative it subsequently.

7. Market and Prices

This section is where appropriate source how the inputs and outputs of project will be treated and indicates what is expected to happen to critical prices and price structure in future.

8. Financial and Economic Implication

The section of financial and economic implication describe the result of financial and economic analysis.

9. Project Justification and Risk

Other impacts and main project list are presented in this section this is often the Ist part of main report to be read by major decision makers hence it is important to concentrate an particular benefits of project rather than results which may be expected.

10. Outstanding Issues and Follow up Options

Finally the last section should draw an attention to any major outstanding issues which must be resolve before project implementation can proceed, thus it also contains suggestion of states to be taken for progress.

Project Appraisal : An Agreement

Project appraisal is normally a process of verification of saturation in the field and security of report documentation. The aim of appraisal are :

1. To evaluate the financial economic and social objective of project.
2. To verify the procedure of project formulation team.
3. Recommend the condition which will ensure that the project objective are held.
4. To ensure that the proposed grant loans expenditure is in accordance with the policy of financial institution.

The farm of appraisal will vary all to the type of project for production oriented project it will normally include the following aspect :

1. **Technical Aspect :** *e.g.,* engineering design and environmental matters.
2. **Financial Aspect :** *e.g.,* Requirement for the funds, the financial saturation of implementing agency.
3. **Commercial Aspect :** *e.g.,* Procurement and marketing arrangement.
4. **Social Aspect :** *e.g.,* Sociological factors and expected impact project of certain gps.
5. **Institutional Aspect :** *e.g.,* Organisation and management arrangements, the requirement of arrangement of technical assistance, project monitoring and revaluation,
6. **Economic Aspect :** *E.g.,* Project cost to the national economy and distribution of benefits.
7. **Tolls for Economic Analysis :** There are number of principles technique used to quantify the profitability of the project based on cost and benefit.
   1. *Pay Back Period (PBP) :* The PBP is defined as number of years required to recover the original investment of the project out of the annual net cash inflow (investment).
   2. *Discounted Catchflow (DCF) :* The Net Present Value (NPV), Internal Rate of Return (IRR), Benefit Cost Ratio (BCR) come under the category of discounted cash flow. The DCF method takes into account the time value of money and it is particularly useful for financing agency the adjustment for time value of money is made through the principles of discounting and compounding.

Note : The method on which the acceptance or rejection of the project is decided or discounting and compounding methods related to accountance.

NPV : The NPV of an investment is the net value of expected future for the calculation of future cost and benefits for the value particular discount rate individual for each project is assumed for which a discount factor is determined the most applicable method for this purpose is consideration of opportunity cost of the capital.

NOH Opportunity Cost : This is the highest return sacrificed by employing resources in particular production process, all to the studies conducted by world bank the opportunity cost of the capital is less developed countries ranges between 6–12%.

IRR : IRR is the project defined as rate of discount which would make the net value of benefits equal to the net value of cost of project.

$$\text{IRR} = r_1 + (r_2 - r_1) \times \frac{P_1}{P_1 - P_2}$$

r_1 = lower discount factor

r_2 = highest discount factor

P_1 = NPV at lower discount factor means NPV RR.

P_2 = NPV at highest discount factor or R_2

If IRR value exceeds the rates of interest the project is accepted.

IRR gives an indication of the rate of interest at which money can be borrowed for financing the project.

B.C.R. (Benefit Cost Ratio)

This ratio of project benefits to project cost which evaluates the efficiency of resources utilization of project BCR is used for :

1. To determine whether the project is to be funded or not.
2. As a criteria for Banking Projects for investment.
3. It is used for computation of economic viability of project.

BCR is computed by using following formula

BCR = NPV of Gross Benefit/NPV of Gross Cost

Sensitivity Analysis

In decision making process particularly in aquaculture the benefit analysis should be followed by risk analysis.

Sensitivity analysis is considered as the Ist step in **Risk Analysis.**

Sensitivity of the project to minor variation in cost and benefits and their impact on profitability of the project has been studies, sensibility analysis has following 3 components :

(*i*) Identification of *Critical Variables* which significantly effect the capital cost at the net cash flow of project.

(*ii*) Specification of alternative value of critical variables relating to cost and return.

(*iii*) Recomputation of NPV and IRR of the project by incorporating alternative values.

Sensitivity analysis helps in understanding all the possible outcomes of investment decision in advance and infacing the financial crisis likely to arise in future.

Project Implementation

There are mainly two aspects of project implementation :

1. Sequence of main implementation phase
2. Problem areas at different stages of implementation

1. Sequence of Main Implementation Phases

(*a*) *Project Phasing :* There are typically upto 6 phases in project implementation, these phases are :

(*i*) Recruiting the human resources

(*ii*) Studies an engineering

(*iii*) Construction and procurement

(*iv*) Startup of field operation

(*v*) Standardization of field operation and achievement of project goal

(*vi*) Termination of project component.

Project need not necessary include all these phases or all in rare integrity *e.g.,* in a mussel culture and marketing project consisting of 4 components a market study for increase production and extension service for few farmers, a credit programme and project management would be necessary also the phases would vary in the length of time they occupy in the project implementation *e.g.* project management will begin when the projects starts and will finish when it ends. While the market study might be fully implemented early in the project schedule. The characteristics of each phase can be described as follows :

1. **Recruiting the Human Resources :** This phase begins as soon as project agreement in signed key steps in this phase of :

    (*i*) Recruiting the project manager and project assistance.

    (*ii*) Establishing the project management office and component and providing them with the means of operation.

    (*iii*) Recruiting project consultant and if needed consultant for specific component. This phase ends when the last recruitment has been made.

Studies and Engineering

This phase begins with the projectment office is operational, key step in this phase.

(*i*) Site Studies and topographic survey.

(*ii*) Monitoring environmental parameters.

(*iii*) Engineering of facilities and preparation of tender documents.

(*iv*) Selection of equipment and supplies and preparation of procurement list.

This phase ends when last construction or supply tender document is fully completed and approved.

Construction and Procurement

This phase begins as soon as the I^st tender and procurement has been approved. Key steps in this phase are :

(*a*) Selection of tenderer or supplier and issue of tender or supplier request.

(*b*) Analysis of offers and awards of contract.

(*c*) Supervision of contractor, workmentor/work management payment of contractor.

(*d*) Work and supply contracts and agreement custom clearness.

(*e*) Post delivery control and guaranties.

This phase ends when the last delivery or construction is completed or receive and terms of guaranties have been met. In construction work this usually means a full year after construction is completed.

(*iv*) Start of Field Operations

It begins as soon as the first field team has been hired and provided with necessary working tools such as vehicles and equipment facilities. Key steps in this phase are :

1. Field testing of all working tools.
2. Meeting expected minimum, biotechnical performance standard.
3. Meeting expected minimum financial performance standard.
4. Meeting expected minimum training, extensions and other standards.

This phase ends when all the field operations has been reached all the expected standards.

(*v*) Standardization of Field Achievement of Project Goals

This phase begins when project start up has been completed and management can turned their attention towards the main project priorities such as selling products from the completed ponds. Improving the performance of project component, this phase with termination of project.

(*vi*) Termination of Project Component

It begins when goals has been reached and purse are exhausted. Key steps are :

1. Preparation for project termination.
2. Project follow-up.
3. Stop realocation.

This phase ends when the last administrative structure has been closed down and all stops disappear.

Budget

Budget is a plan of action or we can say it is a account of income and expenses. Budget is the quantitative statement of the detailed plan and anticipated result of operations during specific period. Budget secure of target to which the operations are directed budget may be physical or financial.

Physical budget covers labour budget, inventory budget and output quantities budget.

Financial budget covers cost budget, capital budget, revenues budget and cash budget, budget in financial terms are conversion of quantity budget ($Q \times M$) monitory budget.

1. Different Points of Project Budgeting

1. Investment or capital budget
2. Inputs budget
3. Liquidity budget (means cash inflow and outflow)
4. Operation budget
1. Invest budget is long-term budget such as 5-year plan budget this involve total outlays of long-term objective, these are also made as milestone of the budget.
2. Budget refers those plans where major long-term durables are creatable such as fish farm cold storage. Fishing boats etc.

2. Input Budget

Budget is also require for purchasing different input items such as fish seed water filling charges, feed etc. it also includes maintenance of capital items and labour charges.

3. Liquidity Budget

This is the most difficult area of Budget since it is the multidimensional factor so less to be successfully estimated and control most of the operational system have period of high inflow and outflow of capital funds *e.g.,* A hatchery has high output during breeding season and high inflow during winter season, control of overflow of the cash is extremely important for maintaining the programme operations and achieving the target within the available resources.

4. Operational Budget

This is the most important point of Budget control in commercial project there are three types of operation budget.

    (*a*)   Production budget which have been supported by raw material.

    (*b*)   Sales budget which has been supported by sold stock.

    (*c*)   Profit budget which has been supported by unsold stock.

These budgets are regulated by 4-central points.

    (*i*) input, (*ii*) cost, (*iii*) time, (*iv*) output

Budget Construction and Control

Budget construction starts with organisational setup of responsibility of function the top management controls the budget through its power to approved it and monitoring progress.

Limitations in Budget

1.   It is stated and not timely and it cannot be changed.

2.   It restrict freedom of management and cannot control waste of initiatively and creativity.

3.   Most serious drawback come from the uniform construction of budget and its thoughtless approach.

Important Features of Governmental Budget

It is also called as DOCTRINE of LADSE Budget sanctioned for one year budget is issued by June, July, September and October revised budget is sanctioned by Feb.-March without leaving any implementing time.

Innovation in Budget

1.   *Flexible Variables :* In this system cost unit output for different level of output are taken into account.

2.   *Zero based Budget :* Past budget is main source of information to decide future budget this is common practice now a days but zero budget is a budget in its own rights not link to the past.

3.   *Programme Budget :* This is the plan of action in a identified frame of time and is more realistic control and progress monitoring is best done in this type of budget so it is mostly adopted well managed private sector.

EXTENSION EDUCATION : CONCEPT AND PRINCIPLES

The word 'Extension' is derived from the lation roots, 'tensio' meaning stretching and 'ex' meaning 'out'. Thus the term 'Extension Education' means that type of education which is 'stretched out' into the villages and field beyond the limits of the schools and colleges. The word 'Extension' the meaning given to it in Webster's dictionary as "branch of a University for students who cannot attend the university proper." Thus, Extension is an out-of-school system of education.

The Basic Concept

The basic concept of extension is that it is education and education is the production of desirable changes in human behaviour – the bringing about of desired changes in knowledge (things known), attitude (things felt), skill (things done), either in all, or one or more of them.

The three kinds of changes in behaviour are illustrated below:

1. *Change in knowledge:* Too heavy application of nitrogen leads to excessive lodging of paddy crop.
2. *Change in attitude:* There is no reason to avoid the application of night soil compost to crops.
3. *Change in skills:* How to prepare good compost.

In some instances, there may be a combination of all the three kinds of changes. For example, in a programme for rat control the extension worker teaches:

(a) That rat can be easily and cheaply controlled by the use of zinc phosphide (Change in knowledge).

(b) That it is no sin to kill rats which are responsible for huge national loss of food stuffs (change in attitude).

(c) How to use zinc phosphide as a poison bait, without undue risk (change in skill).

In education, emphasis is usually placed on changes in knowledge and skill, more or less neglecting the aspect of attitudes. But attitudes (or emotions or feelings) are important, because they tend to express themselves in action which may be favourable (positive) or unfavourable (negative) to public interests or progress.

Concept of Extension Education Process

This concept identifies five essential phases in Extension Education Process. Figure 1 shows the sequence of steps in a cycle that may be expected to result in progress from a given situation to a new desirable situation.

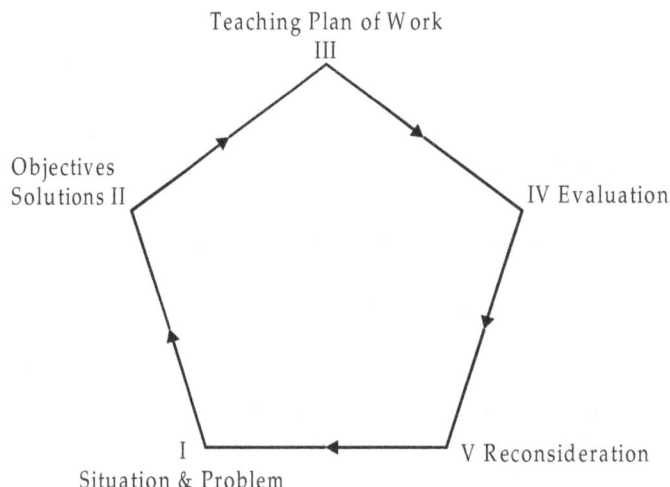

Fig. 1 : The Extension Education Process.

(*i*) The first phase analysis of situation requires *collection of facts* about all aspects of situation. Facts are needed about the people, their interest, *education, habits* etc. Facts are needed about physical situation such as soil, type of farming, communication etc. In real life situation, the farmers are not in a position to continue use of all the practices recommended to them and there is some gap. This analysis will examine change conditions and take a careful look ahead, comparing "*What is*" with "*What should be*".

(*ii*) The second is deciding upon objectives. It is necessary to select limited number of problems and to state their objectives clearly. The solutions to be offered must give satisfaction.

(*iii*) The third phase is teaching. This involves choosing (*i*) the content or what is to be taught and (*ii*) Methods, the techniques of communication. The ability to choose and use those methods best adopted to particular objectives is the measures of an extension worker's effectiveness.

(*iv*) The fourth phase is evaluation of the teaching. This should determine to what extent the objectives have been reached. This will also be a test of how accurately and clearly the objectives have been stated.

(*v*) The fifth phase is a reconsideration after evaluation has been taken place. This step consists of a review of previous efforts and results which reveal a *new situation*. Then the whole of this new situation shows the need for further work, hence this process is continuous.

Principles of Extension

Principles are the generalized guidelines which form the basis for decision and action in a consistent way. The universal truth in extension which have been observed and found to hold good under varying conditions and circumstances are enumerated below:

1. **Principle of cultural difference:** Culture simple means social heritage. There is cultural difference between the extension workers and the farmers. Differences exist between groups of farmers also. The difference may be in their habits, customs, values, attitudes and way of life. Extension work, to be successful, must be carried out in harmony with the cultural pattern of the people.

2. **Grass-roots principle:** Extension programmes should start with local groups, local situation and local problems. It must fit to the local conditions. Extension work should start with where people are and what they have. Change should start from the existing situation.

3. **Principle of interests and needs:** People's interests and people's needs are the starting points of extension work. To identify the real needs and interests of the people are challenging tasks. The extension worker should not pass on his own needs and interests as those of the people. Extension work shall be successful only when it is based on the interests and needs of the people as they see them.

4. **Principle of learning be doing:** Learning remains far from perfect, unless people get involved in actually doing the work. Learning by doing is most effective in changing people's behaviour. This develops confidence as it involves maximum number of sensory organs. People should learn *what to do, why to do, how to do* and with what result.

5. **Principle of participation:** Most people of the village community should willingly cooperate and participate in identifying the problem, planning of projects for solving the problems and implementing the problem in getting the desired results. It has been

the experience of many countries that people become dynamic if they take decisions concerning their own affairs, exercise responsibility for, and are helped to carry out projects in their own areas.

The participation of the people is of fundamental importance for the success of an extension programme. People must share in developing the programme and feel that it is their own programme.

6. **Family Principle:** Family is the primary unit of society. The target for extension work should, therefore, be the family. That is, developing the family as a whole, economically and socially. Not only the farmers, the farm women and farm youth are also to be involved in extension programmes.

7. **Principle of leadership:** Identifying different types of leaders and working through them is essential in extension. Local leaders are the custodians of local thought and action. The involvement of local leaders and legitimization by them are essential for the success of a programme.

   Leadership traits are to be developed in people so that they of their own shall seek change from less desirable situation. The leaders may be trained and developed to act as carriers of change in the villages.

8. **Principle of adaptability:** Extension work and extension teaching methods must be flexible and adapted to suit the local conditions. This is necessary because the people, their situation, their resources and constraints vary from place-to-place and time-to-time.

9. **Principle of satisfaction:** The end product of extension work should produce satisfying results for the people. Satisfying result reinforce learning and motivate people to seek further improvement.

10. **Principle of evaluation:** Evaluation prevents stagnation. There should be a continuous built-in-method of finding out the extent to which the results obtained are in agreement with the objective fixed earlier. Evaluation should indicate the gaps and steps to be taken for further improvement.

EFFECTIVE TRANSFER OF TECHNOLOGIES

Presentation of information is equally important as its possession. Percentage of adoption depends on how it is presented. In field of agriculture many technologies were generated but a few could reach to doorsteps of farmers. While delivering goods, the extension personnel are required to communicate technologies effectively. It is made easy it they upgrade their presentation skills. Irwing Wallace reports that number one fear before speaker is to communicate of speak before a group of audience how knowledgeable they may be. Modern information and communication technologies and good presentation skills can help in capacity building of extension personnel in this regard.

Presentation is "Suggestion of an idea as a solution to any given problem". To create, develop and deliver effective presentations is a learned skill as no one is a born speaker. In presentation situation three important elements *viz.*, speaker speech and audience constitutes speech triangle.

Extension Officer (Speaker)

Speech Triangle

(Content)
Improved Agril.
Fishery Technology

Farmers
(Fisher folks)

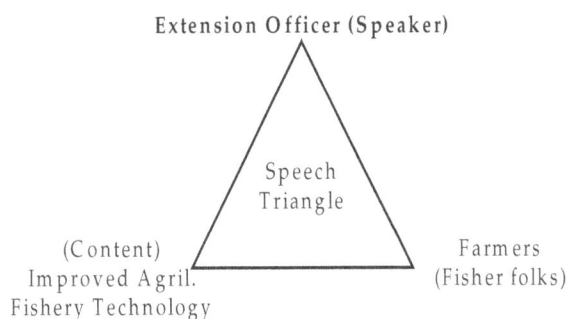

Basic skills needed for extension officer for making effective presentations are :

1. Creative/developing skills
2. Delivering skills
3. Feedback skills

Creative/Developing Skills: This involves planning, preparation, rehearsal and speaker considerations.

Planning Impromptu Talks/Presentations

Audience analysis, situation analysis and goal and aim analysis will help in systematic planning. The following four methods can be used for planning considering the situation, time, audience and subject matter.

PREP Method: PREP stands for point, reason, example and point.

Point : Consider personal view point, something strong you feel and what is the attention getter in the topic/issue.

Reason : Explain why you feel this way.

Example : Quote real life examples which clearly illustrate your view and support your material.

Point : Go back and again restate the point for more impact.

Past/present/future method: Get what happened in the past? (past), What is the present situation? (Present), Where are we going? (Future). If these questions about specific issue/aspect are answered, it means you have planned.

Related incident method: Use the subject as reminder of the previous incident. Relate the incident or the previous experience using lot of details and examples. Ex: If you have to speak on use of plant protection equipment, tell your experience of story, use a role play show how hazardous it is relating it, to an experience you have come acrossed.

Five 'W's & one 'H': By obtaining answer for the following questions one can very well plan for presentation.

What?	Happened
	Was the cause
	Is responsible
Who	Is involved
Where?	Did it happen
When?	Did it happen
	Did it happened
Why?	Are you involved
	Did it happened
How?	Are you involved

By using any one of these methods presentation can be effectively and easily planned.

Preparation: "To fail to prepare is to prepare to fail". Collect all related information from books, journals, expert's views and experiences. Write down all possible ideas, information and experiences related to topic and mull over the theme through out. Use mind mapping for developing content. Ensure that the body of informations has essentials, (aspects that audience must know) desirables (should know) and possible (could know), so that you can adjust/edit content depending on the time available.

Building the content for presentation demands certain skills like humanizing, personalizing, visualising specifying and dramatizing the information gathered. Collected information may be divided into

1. Introduction
2. Body of information
3. Summary & conclusion and
4. Feedback

Prepare answers for all possible questions you may face during presentation.

Visuals Preparation: Visual aids you use must be relevant, appropriate and support your presentation. Ex: charts, flash cards, specimens, etc. To make your presentation effective, power point can be used in which you can also use animation, video clippings of success stories of farmers, real field situations, etc.

1. Arrange or keep ready all the visual material and equipment ready in sequence for your presentation.

Speaker Preparation Through Rehearsal: Charless F. Kittering (US speaker) Said "I hardly believe in preferring to write on piece of paper, rather I like to prefer to write on minds of audience; Piece of paper cannot stand between me and my people". This expounds the importance of speaker preparation through rehearsal. Present before a mirror or your colleagues and check for timing, sequential and appropriate use of visual materials, equipment and your voice and non-verbal skills.

Just Before Presentation

1. Take deep breath 1 to 10 times before you get on to stage.
2. Have pep prepared talk-that you would present well.
3. Keep mind occupied.

Presentation Skills Required While Delivering the Talk

Introduction: Critical part of presentation is introduction as the speaker is judged with in first 60 seconds.

1. Mind, body and voice of the speaker must energise the presentation.
2. Give personal introduction. Don't apologise in introduction for any reason.
3. Use zinger's to capture audience attention like-Key notes, vital statistics, quotes of famous people, arresting titles, importance of the issue or success story, analogies and anecdotes, etc.
4. Allot 15 per cent of total time of presentation for introduction.
5. Talk @150 words per minute.
6. Use a poster, specimen, present a dialogue or crack related joke to sustain the interest of audience.

Body of Information

1. Allot 70-80 per cent of time of total presentation.
2. Present only key essential (must know) points with support from visuals.
3. If time permits cover desirables (should know) and on further queries form audience cover possible (could know) also.
4. State emphatically how problem can be solved with provided information.
5. For any question don't give your personal opinion but present what your organization wants.
6. Repeat, rephase the question before you answer.
7. Tackle if syndrome people by using humour and make them realize that too many fish become more hypothetical.
8. Don't speak to the visuals (chart, poster, power point slides, etc.)
9. Use visuals in the sequence properly as you have planned.
10. Make it interactive to avoid monotony.
11. Laughter is no enemy to learning says Walt Disney so add humour where ever relevant.

Summary and Conclusion

1. Allot 10 per cent of total time of presentation for summarizing and concluding.
2. Recapitulate is short all the essential points emphatically.
3. Climax presentation and conclude with an appeal and drive for action.
4. Inform about your follow up activities.
5. Conclusion is most important as last impression could last long in minds of people.

Feedback and Follow up Skills

1. Invite question for better participation and to see how far the message has reached audience mind. For this use direct, open, general and passed on questions to audiences as per your plan.
2. Conclude the section with an appeal for action. Return to the final there of presentation and thank the audience.

Non-verbal Communication Skills Required for Effective Presentations

The discovery of non-verbal communication has transferred the study of human social behaviour–according to Bird Wistell (1977) 65 per cent of communication is through non-verbal communication while 35 per cent is only communicated verbally. Non-verbal communication refers to all external stimuli other than spoken or written words comprising appearance, para-language, proximics and kinesics.

Appearance

1. Dress neatly and tidily; because when there is no record of behaviour of a person, people tend to judge by their physical appearance.
2. Carry your self in a confident and professional manner.
3. Avoid too many artifacts.

Paralanguage : (Voice Skills)

Voice Quality : Pitch, resonance, volume, rate and rhythm determines the quality. "Monotonous voice is a liability to the audience."

Voice Characteristics : Laughter, coughing throat clearing and sighing frequently may be avoided.

Vocal Qualifies : Variations in pitch and volume. When you emphasize, dramatize or use humor, variation in pitch and volume is needed.

Vocal Segregates : Silent sounds such as 'has', ars' and pauses. Avoid vocal segregates unless when emphasized or dramatized.

Use "PAMPERS" for Effective Presentation

Projection : See that your voice reaches last row.

Articulation : Speak words clearly and distinctly without trailing of end of words/sentences.

Modulation : Use high and low pitch for emphasis.

Pronunciation : Pronounce words clearly.

Enunciation : Over emphasize, accentuate syllables.

Repetition : Repeat as "Repetition is mother of retention".

Speed : Change the speed when you use humour, dramatize or emphasize.

Proximics : It is study of interpersonal distance there are four zones of interpersonal distance *viz.,* intimate, personal, social and public. It is appropriate to use 4'-12' distance (social zone) during presentations.

Kinesics : Study of physical movement in body. Most of the human communication takes place be use of gestures, postures, position and facial expression.

Have pride in your posture and movement

1. Be natural-don't move around too much or too little.
2. Move forward for emphasis.
3. Move diagonally or side to side to engage all parts of the class avoid barriers, stand square.

Eye Contact

1. Keep eye contact with the audience.
2. Don't stare or intimidate.
3. Don't look out through window or at clock or ceiling or at your feet.

Gestures

1. Use meaningful and appropriate gestures to make a point.
2. Don't play with keys, chalk, coins, etc.
3. Don't use hands too much touching your nose, forehead and ears.
4. Avoid foot tapping, scratching head.
5. Avoid pens or pencils for pointing any body in-group.

6. Older audience seek restrained gestures.

7. Avoid repetition of favourite gestures.

Mannerisms

1. Don't tap microphone

2. Avoid hands in pocket all the time.

3. Avoid holding a pointer or chalk through out.

Conclusion

Extension personnel at different levels by and large are regularly involved in training programmes for imparting knowledge and skills to their lower people and ultimately to farming community. It they upgrade their creative/developing, delivering skills they can present the subject matter in an interesting and understandable manner more effectively. Hence training institutes of extension functionaries should concentrate efforts to improve or upgrade their communication skills in particular; which not only facilitate in qualitative training but also the trainers for their self improvement (AER).

Chapter 2

COMMUNICATIONS

India has made giant strides in the field of space research and nuclear power in the last few decades. Just the other day there were talks about EU including India in its Galileo global positioning system and its energy panel. But does it mean anything to the average Indian? Do ordinary denizens have the faintest inkling of his or her country's scientific development? What does the word "science" mean? Ironically, the latin origin of the word means to know. But what is the need to know? And why do we need scientific awareness? How can it contribute to the progress of the country?

It is a known fact that there are several wrong notions about everyday events and happenings in our country that can be directly attributed to ignorance on a large scale. To make any kind of progress in any direction the country has to remove such public perceptions which are nothing more than mere superstitions. And the only way to do this was to take the faculty of scientific thought to the masses so that their understanding of things are based on intelligent reasoning and not heresy or make-believe.

So it was only imperative that 2004 was declared the Year of Scientific Awareness (YSA).

The idea of observing 2004 as the Year of Scientific Awareness was conceptualised in a meeting of representatives of state science & technology councils held in Technology Bhavan, New Delhi, while discussing plans of science communication activities for the 10th Five Year Plan. The first announcement to this effect was made at the inaugural function of 10th National Children's Science Congress held at Mysore during December 27-31, 2002. YSA-2004 was formally launched during the inaugural session of the Indian Science Congress Association (ISCA), held from January 3-7, 2004 at Chandigarh.

The programmes seek to:

1. Make as many people scientifically aware as possible.
2. Make more and more people habitual of keeping themselves scientifically aware by acquiring the required knowledge and information and seeking satisfactory answers to questions that arise in their minds.
3. Help create an atmosphere and conditions conducive to more and more people readily and easily becoming scientifically aware.
4. Encourage more and more people to make practical use of their scientific awareness in day-to-day life, in arriving at decisions concerning issues/subjects of concern to them, in overcoming superstitions and tackling blind beliefs, and in handling situations arising out of age-old practices and traditions which actually hinder progress, harmony, or even may harm others.

The National Council for Science and Technology Communication (NCSTC) or Rashtriya Vigyan Evam Prodyogiki Sanchar Parishad (RVPSP), Department of Science and Technology,

Government of India has, therefore, taken off from its past efforts and has engaged in channelising their efforts in a joint and sustained manner to take science awareness to every corner of the country, explains Engineer Anuj Sinha, scientist and head of the organisation.

YSA-2004 is a platform for the people of India to come forward, work and move together and make the dream of modern India come true scientifically. It is hoped that the inculcation of scientific spirit and bent of mind would metamorphose developing India into a developed one. The scientific approach is to be reflected in the entire cross section of people of our country by which the dream would be fruitful.

The YSA-2004 was conceived as a campaign of one full year of multiple level activities that have been conducted across the length and breadth of the country. There have been two type of activities-jatha and non-jatha.

Jatha : The word jatha immediately brings to mind long rallies that often are the cause for bringing life to a standstill particularly in cities. But here are jathas that people would be glad about. The Vigyan Chetna jathas have been providing educational and motivational entertainment. The programmes have been conducted at pre-planned 'halts' of the jatha troupes. All the activities have been built around selected issues of major concern and involve two-way interaction with audiences.

It has turned out to be very useful and effective means of reaching the masses. This interactive mode of communication is found to have better impact than most other methods. Using the technology of 'jatha', RVPSP has been successful in spreading the message of science and technology in the Bharat Jan Vigyan Jatha-1987 and Bharat Jan Gyan Vigyan Jatha-1992.

Vigyan jathas comprise groups of people well versed in different communication formats, possessing multimedia skills, who move together on a well-defined route to communicate messages of science and technology, with activities or software built around a focal theme. At the helm of affairs is the National Organising Committee with the country's best minds as its members. Dr Vasant Gowariker, former scientific adviser to the PM is the NOC'S president with UNESCO Kalinga prize-winner Dr Narender K. Sehgal as its chairman.

The Vigyan Chetna jathas have been focusing on various local issues in different states. Dry runs or discussion meetings with prominent community groups were organised specifically to identify local needs. The Vigyan jathas are not forced upon people as a surprise. They are well planned and pre-jatha activities by identified local groups give the required pre-publicity to them. YSA-2004 has been planned and have conducted jathas across 80 to 100 km wide belts in each region.

The Vigyan Chetna Jathas are issue based and their routes pre-defined. The Local Organising Committees (LOCS) make local arrangements like publicity, media coverage etc. under the supervision of District Organising Committees (DOCS). The Jatha group consists of 10 to 15 performers (local artists for folk as well as street plays etc.), 3 to 4 volunteers for managing the performance along with expert scientists for addressing the local issues. They move on pre-defined routes on well-decorated vehicles that carry material and other utility items.

The stoppages have been planned after every 25 or 30 km with night-halts at main venues and 3-4 performances at each place. Public performances have been held in the evenings. Depending on the location, jatha troupes perform at different nearby/surrounding locations too. Each group of artists work for 10 to 12 days with 4-5 experts or scientists who are from the same or belong to nearby districts or localities.

With its main focus on creating awareness on the key areas of concern taken up by the YSA-2004, the jathas have been focussing mainly on health and nutrition, soil cover management, water and sanitation, disaster preparedness, empowerment through IT, environment and conservation of bio-diversity.

Non-Jatha Activities : Non-jatha programmes meant organising round-the-year activities in academic institutions, research organisations, associations of professionals and groups of people working together. Even though the basic concerns of these activities are the same as Jatha concerns, the activities here will be relatively for larger groups and more focused towards specific end. The various target groups and different activities planned are:

Schools : Hands-on activities like water testing, waste management, checking food adulteration, nature camps, understanding plants, learning science by doing, quiz and other competitions, visits to research and development labs and interaction with scientists.

Organised sectors like factory workers **and NSS members :** Activities among them would mainly revolve around lectures and debates.

Gram Panchayats : Sensitisation on applications of IT on issues like e-post, maintaining land records, disaster preparedness, consumer awareness, nutrition and health.

General Masses : Exhibitions, Science Fairs, Rallies, Folk performances, etc.

The activities will be carried in 30 states and union territories covering nearly 600 districts. To organise year long programme effectively the whole country has been divided into specific regions with set of states having similar geographic features and problems.

Vigyan Rail : This is one of the most imaginative of activities that has been undertaken under the science awareness programme.

"Vigyan Rail—Science Exhibition on Wheels", conceived, formulated and implemented by Vigyan Prasar, an autonomous body under the Department of Science and Technology, jointly with the Ministry of Railways was truly a unique project. It has been undertaken with the active participation of scientific departments/ministries/councils of the government. Vigyan Prasar jointly with National Council of Science Museums prepared a detailed project report. To discuss the various aspects related to the Vigyan Rail Project, and to seek the views, suggestions and commitment to the project, a coordination meeting of scientific departments/ministries/councils was held in July 2003. The response was truly overwhelming. The costs were being shared by the DST, Vigyan Prasar, and other participating departments/ministries.

Vigyan Rail displayed exhibits/activities depicting India's achievements in various fields of science and technology. The exhibits were mainly based on India's scientific heritage (National Council of Science Museums), environment (Ministry of Environment and Forests), space (ISRO), communication (Department of Communication), information technology (Department of Information Technology), scientific and industrial research (Council of Scientific and Industrial Research), ocean development (Department of Ocean Development), water resources (Ministry of Water Resources), defence (Defence Research and Development Organisation), agriculture (Indian Council of Agricultural Research), non-conventional energy sources (Ministry of Non-Conventional Energy Sources), Health and medicine (Indian Council of Medical Research), atomic energy (Department of Atomic Energy), meteorology (India Meteorological Department) and survey and mapping (Survey of India).

Vigyan Prasar and Technology Information, Forecasting and Assessment Council (TIFAC) and organisations under the department also participated with their contributions. In about 200 days, the exhibition visited 57 towns and cities allover the country drawing lakhs of visitors.

The exhibition, after completing its first phase, returned to Delhi in August 2004. It is being refurbished for a second run to take the exhibition to another 25 towns.

Information Technology : There is no doubt that tomorrow's world hinges largely on information technology. So any distance from it means being alienated from the world, living in isolation. It is a skill that has to be acquired like any other subject: English, Science and Mathematics. The importance of knowing the basic of computers has caught the attention of all without any divide. Digital divide is the new terminology for another of the growing inequities in a world of globalisation. Computer education is being made available as part of school curriculum all over India and being brought to the masses. Andhra Pradesh, Karnataka, Kerala, Tamil Nadu and West Bengal have already implemented large projects in this area while others are likely to follow suit soon.

Over 20 lakh school children are currently getting exposed to computer knowledge and this number is only going to increase with more states taking the initiative toward bridging the digital divide. It is a well recognised fact that Internet-based education is a cost effective and efficient way of reaching out to learners across the globe, even those in remote areas.

The increasing use of Internet has caused a paradigm shift in the mode of training delivery. IT boom in India is filtering down through little-known, less glamorous programmes to far-flung areas and transforming the backward classes of our society.

One good example of how IT is an important component of science awareness activities are the programmes being put in place in the state of Andhra Pradesh. It illustrates the extent to which not just literature on science but information technology has been taken to the masses. The committee responsible for it identified computer software; besides, 50 sets of books and two sets of posters have been purchased and given to each district: Booklets on Malaria, Anaemia, Safe water, Vitamin A deficiency, Water-borne diseases, Constructing lavatories, About breast feeding, Nutritional receipes, Conserving nutrients while cooking, About complementary feeding, Personal hygiene, Disaster preparedness (cyclones) were brought out.

Posters sets (10-15 posters) on different health topics like family planning, infectious diseases, community environment and personal cleanliness, healthcare for pregnant women, child healthcare, health care for women, first aid and nutritious food were printed. The software that is being prepared would delve on topics like water, understanding natural and man-made disasters, tackling disasters, design of a non-computerised search engine.

Introducing planning tools based on IT to the panchayat members is being attempted under YSA-2004 in selected blocks. The decision-making will witness a major improvement after this project.

Empowerment : One of the significant achievements of YSA-2004 would be the way it would have made Indian women self-dependent. Rural women in India who have little say in their own lives and their families, have now changed albeit slowly. The change is visible in many subtle ways like where village women in many areas have done away with corrupt moneylenders who fleece the illiterate. Women self-help groups now save money in banks and keep track of their savings on computers. Access and control of information, however small it maybe, by the poor and downtrodden of the country is a big step towards empowerment.

But 2004 will not be curtains for the department of science & technology's efforts to create scientific awareness in the country. DST is extending it to the future too.

Professor V.S. Ramamurthi, Secretary to the Department of Science and Technology in the Government of India said: "There are two or three things which we definitely look forward to.

One is an increasing number of science communicators who will convey the developments in S&T to the people, even if they are not professionals in the field of communication. For instance, nothing stops a college teacher from writing an article every month on an issue of public importance. His words find echo and enthusiasm in Er. Anuj Sinha, head of DST'S Science Communication department, who is the spearhead of the department's activities for the future. Which is why scientific awareness in the department's hands have become a larger process to make people participate in a continuing endeavour of scientific self-realisation to build a modern, sustainable and developed India.

Chapter 3

COMMUNICATION PROCESS

1. Introduction

❑ As we know communication is – dynamic, ever changing and unending (no beginning or end). How do we get hold of it. Although task is not easy it is not impossible.

❑ As we know "Communication is a process that involves an interrelated, interdependent group of elements working together as a whole to achieve a desired goal/outcome".

❑ "Communication is a process, whereby two or more people exchange ideas/facts/feelings is a way that there can be a common understanding to each other." J.P. Legans.

❑ "Communication is nothing, but sharing of experience" Wilber Shram.

❑ So we can study communication in much the same way we study biological system/process with in our body (digestive circulatory system). We can determine the involved, analyze how these elements affect one another and this determines the nature of process as a whole.

2. Characteristics of Communication

"Communication is a process in which inter-related elements work together to achieve a desired goal/outcome". As we view communication is a process, so we also perceive it to be-

- ◆ Dynamic
- ◆ Ever changing
- ◆ Unending

❑ Additionally thousands of information, ideas, opinions, that you process, evaluate and store each day also change you to some extent. So every next moment you are a changed person.

❑ Another key concept of communication – communication events don't occur in isolation from one another. Each interaction/communication that you have affects, each one that follows is not always in simple and direct manner.

❑ 70% of communication is non-verbal and 30% is verbal.

3. Dimensions of Communication

Communication is more than one person speaking and another listening. This section will explain some of the complexities of communication by looking at the dimension of :

❑ Verbal and non-verbal communication

❑ Oral and written communication

❑ Formal and informal communication

❑ Intentional and unintentional information

In addition two indirectly dimensions – Human-computer communication and Animal communication.

(I) Verbal Communication

"Verbal communication involves the use of symbols that generally have universal meaning for all who are taking part in the process."

❒ As such verbal communication may be spoken and spoken.

❒ These spoken/written symbols are known as language

❒ Additionally verbal communication is highly structured and use formal rules of grammar.

(II) Non-Verbal Communication

Involves the use of symbols other than the written and spoken words such as :

- gestures
- eye behaviour
- tone of voice
- use of space
- and touch etc.

❒ Although non-verbal communication have social-shared meanings, they have no formal structure and grammar rules.

❒ Non-verbal communication usually complete/support the verbal communication.

❒ At other time non-verbal symbols completely replace verbal-messages.

(III) Oral Communication

Refers to the message that are transmitted out-loud from one person to another. Most messages are verbal, with complementary non-verbal messages. Each day we participate in Oral communication either as a speaker or listener.

(VI) Written Communication

It is primarily verbal communication but non-verbal characteristics can effect it. Written communication is taking place right known when we are reading this book, because this is a writen document.

Oral communication and written communication serve different functions and are used both independently and in combination.

(V) Formal Communication

In formal communication such as public speaking or mass communication, we pay more attention to both verbal and non-verbal messages.

❒ Language use is more precise, with careful attention to grammar.

❒ More concerned on non-verbal items such as dress, postures or eye contact.

❒ Formal communication is also used with person of higher perceived status.

(*VI*) Informal Communication

Such as inter personal and small group communication people are at ease.

❐ Observes would notice more hesitation and slang in verbal messages.

❐ Less attention to non-verbal messages like clothing, postures, eye contact etc.

(*VII*) Intentional Communication

❐ Intentional communication occur when messages are sent with specific goals in mind.

❐ Well planned, well designed, mix both verbal and non-verbal with target audience in mind.

(*VIII*) Unintentional Communication

❐ When communication takes place unintentionally without the communicator's being aware of it.

❐ The greatest number of unintentional messages are non-verbal. Our non-verbal behaviour often speak louder than words.

(*IX*) Additional Communication forms

❐ Two additional communications forms are rapidly becoming of interest to communication scholar:

(*a*) **Human – computer communication:** Computer technology has established its prominence in this age of information. Now scholars are interested in the effects of technology on human communication, *e.g.*

♦ To what extent computer mediated communication depersonalize interaction.

♦ What effects do computerized conferences have on the group decision making process.

♦ To what extent does computer mediation reduce/increase communication efficiency.

(*b*) **Animal communication:** Specific researches have shown that animal share with human a number of characteristics including those associated with–attraction and making, territoriality, rivalry, play, family ties, familial ties, colonial organization, division of labour and other habits that we once assumed were uniquely 'human'.

❐ In short, like human, animals use wide variety of communication process ranging from

♦ Auditory

♦ Visual and tactical system to olfactory.

♦ Thermal and

♦ Electrical system.

❐ With hope, in future, scholar will answer the questions related 'language pattern of animals'.

4. Communication Context

❐ We communicate in different context like :

♦ Intra personal communication (communication with self)

♦ Inter personal (communication between two).

♦ Small group communication (communication between 5-7).

- Organized communication (communication between an organization).
- Public communication (communication between leader and public)
- Mass communication (communication with infinite number of people)

5. Communication Process

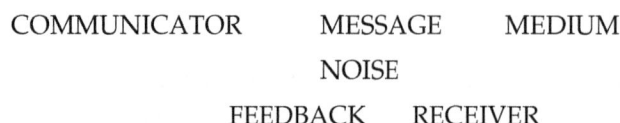

COMMUNICATOR   MESSAGE   MEDIUM

NOISE

FEEDBACK   RECEIVER

6. Elements of Communication

❑ Applying this approach to communication process we find seven elements :

❑ A source/encoder/communicator which send *A message* through a channel to a *receiver/ decoder* which responds via *feedback* with possibilities of communication break down in each stage *i.e Barriers/Noise*. However, none of these elements is meaningful outside of a Situation/context in which it may be interpreted.

(I) *The Source (Encoder)*

❑ The source makes the decision to communicate.

❑ It determines what the purpose of message will be :

- To inform
- To persuade
- To entertain

❑ First the message generates through source – past experience, Perception, Thoughts and Feelings.

❑ Then source encode the message (give it a form which can be communicated/ communicable), and form Verbal, Non-verbal Written or Spoken.

Characteristics of a Good/Effective Source

- Knowledge of (subject)
- Credibility (Trust, expertness and experiences)
- **Role perception** (actual presentation, if possible)
- **Communication skills** (clear, brief, non-verbal, body language)
- **Attitude** (Positive and Negative).
- **Compatibility** (in terms of language, skill)

(II) The Message

(*i.e.* the information which is being communicated)

❑ Source after deciding, what message will be transmitted, the source uses symbols to get across to others, i.e. (words, pictures, visuals etc.) *i.e.* decode it.

❑ A message is given a language (verbal/non-verbal).

Characteristics of a Good Message

- Communicable

+ Understandable
+ Profitable
+ Divisible
+ Simple (made simple)
+ Need based/timely
+ Attractive

(III) Channels

❏ Channels are means (pathway/devices) by which messages are communicated.

❏ Channels may be described and analyzed in two different ways-

(A) The form in which messages are transmitted to receivers *i.e.*

+ Verbal channels of communication.
+ Non-verbal channels of communication.

(B) Channels may also be described according to manner of presentation

+ Face to face (Interpersonal communication channel)
+ Use of public address system (Public communication channel)
+ Talk over radio/television/newspaper (Mass communication channel)

Characteristics of Goods Channel

❏ Attractive
❏ Economical
❏ According to audience (number, audience profile)
❏ According to time and space
❏ Multi-sensory
❏ Manageable
❏ Brief
❏ Accurate
❏ Modern

(IV) Receiver/Decoder (*audience*)

❏ The person(s) who attends to/receive the sources's message is the receiver.

❏ The act of interpreting messages is called decoding, one who performs decoding is decoder.

❏ Receiver decodes messages based upon his Past experience, Perceptions, Thoughts and Feelings.

❏ We receive messages through all our senses. But mostly we decode message by listening/seeing.

Characteristics of a Good Receiver

❏ Ability
❏ Interest

❑ Background

❑ Custom/beliefs

❑ Confidence in source

❑ Social status of source

(V) Feedback

❐ Each party in an interaction continuously sends messages back to other. This return process is called feedback. Feedback tells the source, how the receiver has interpreted his/her message?

❐ Feedback are mainly of five types :

(*a*) Positive feedback

(*b*) Negative feedback

(*c*) Ambiguous feedback

(*d*) Direct feedback

(*e*) Indirect feedback

(VI) Barriers

As in radio/T.V. transmission distortion may occur at any time. Similarly there may be noise/burner at any stage in communication process.

❑ It may be at source (language/voice/tone etc.).

❑ It may be thought channel (wrong selection of channel)

❑ It may be in message (language/format etc.)

❑ It may be in decoding/receiver level.

❑ Even in feedback of receiver to source.

❑ If sender/receiver are not on same 'wavelength' (experience level)

❑ Mechanical noise as well.

(VII) Context/Situation

This element of communication is, perhaps the most important, it effects each element of the communication and communication process as a whole. So always keep in mind who says, To whom, With what objective/intention in mind and in what situation/environment/context.

AUDIO-VISUAL AIDS

1. Introduction

Audio Aids : The instructional devices through which the messages can only be heard are known as Audio aids.

Visual Aids : The instructional devices through which the messages can only be seen are known as Visual aids.

Audio-visual Aids : The instructional through devices which the messages can be heard and seen simultaneously are known as Audio-visual aids.

Or you may say : **'The audio-visual aids are instructional devices which are used to communicate messages more effectively through sound and visuals.'**

2. Why Audio-Visual Aids?

The use of audio-visual aids has following advantages:

- ❑ Capture audience attention and arouse their interest.
- ❑ Highlight main points of message clearly.
- ❑ The possibility of misinterpreting concept is reduced.
- ❑ Structure the learning process more effectively.
- ❑ Help reach more people irrespective of their level of literacy and language.
- ❑ Speed up process of learning and teaching by supplementing and supporting.
- ❑ Save time of the teacher and the listener.
- ❑ It activates and motivates the lister/audience.
- ❑ It summarizes and emphasize better.

Messages perceived with serval senses are understood and retained better, *viz.*

Mode of Instruction	Retention Level
Reading	10%
Hearing	20%
Seeing	30%
Hearing and seeing	50%
Doing	70%
Repeated doing	90%

3. Classification of Audio-Visual Aids

Audio-visual aids may be classified into following categories:

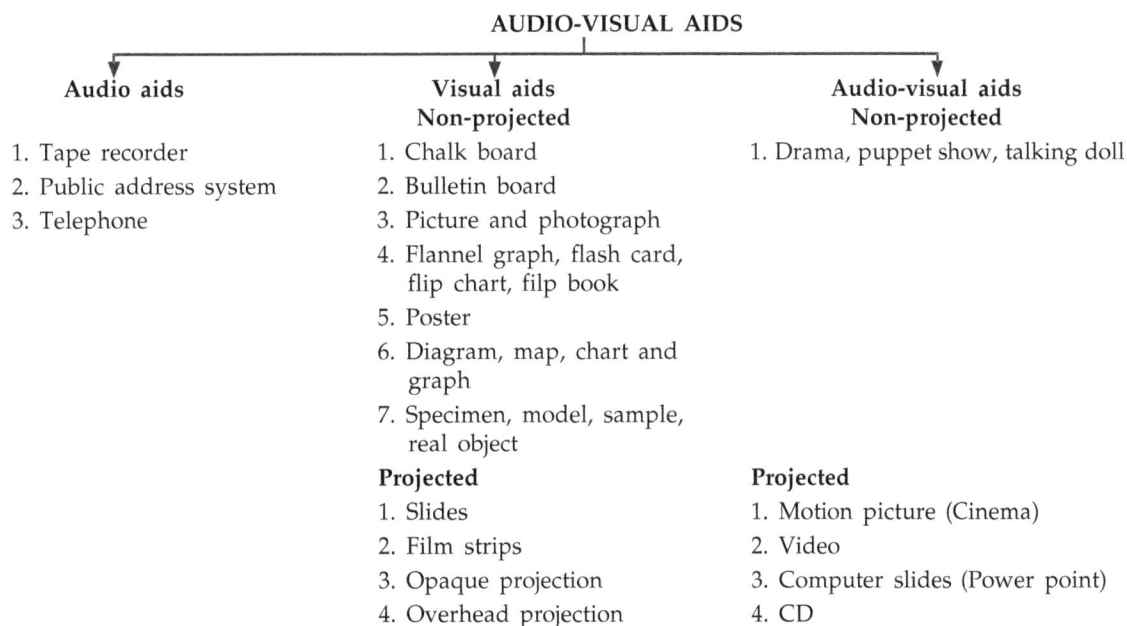

AUDIO-VISUAL AIDS

Audio aids	Visual aids Non-projected	Audio-visual aids Non-projected
1. Tape recorder	1. Chalk board	1. Drama, puppet show, talking doll
2. Public address system	2. Bulletin board	
3. Telephone	3. Picture and photograph	
	4. Flannel graph, flash card, flip chart, filp book	
	5. Poster	
	6. Diagram, map, chart and graph	
	7. Specimen, model, sample, real object	
	Projected	**Projected**
	1. Slides	1. Motion picture (Cinema)
	2. Film strips	2. Video
	3. Opaque projection	3. Computer slides (Power point)
	4. Overhead projection	4. CD

4. Choice of Audio-Visual Aids

The choice of audio-visual aids shall depend on number of criteria.

  i. **Teaching objective:** Whether to give information, to impart skill or to bring change in attitude etc.

 ii. **Nature of subject being taught:** Particular aspect of the technology and whether understanding depends on seeing or not.

iii. **Nature of audience:** Their age education, interest, experience, knowledge, intelligence etc.

 iv. **Size of audience:** Small or large

  v. **Availability of equipments,** materials and funds

 vi. **Skill and experience** of the extension agent in preparation and use of audio-visual aids.

5. Organization of Audio-Visual Programme

The following general procedure may be adopted in organizing an audio-visual programme.

A. Planning

- ❑ Identify the objective.
- ❑ Plan for a simple, practical, educational and interesting presentation.
- ❑ Anticipate the size of the audience.
- ❑ Plan for a variety of colourful visual aids. It includes :
  - ✦ Develop talk plan
  - ✦ Identify essential points
  - ✦ Decide form of visual
  - ✦ Rough in design and layout

B. Preparation

- ❑ Collect relevant equipments and materials.
- ❑ Compose contents
- ❑ Pre-view the audio-visual aids.
- ❑ Finalize the audio-visual aid and make ready for use.
- ❑ Check power supply, lighting, need for total darkness, seating arrangements etc.
- ❑ Select and train audio-visual operator or any suitable person in the organization.
- ❑ Arrange audio-visual aids in a sequence and have them within easy reach.
- ❑ Few guidelines for preparing visual-aids :
  - ◆ *For written form :*
    Use familiar words
    Write capital letters
    Uniform height
    Similar gap
    Bold letters
    Uniform slant

◆ *For illustration :*
   Simple and direct
   Large
◆ *General :*
   Brief
   Balance
   Accurate

C. *Presentation*

❑ Motivate the audience and stress on the key points.
❑ Present aids at the right moment and in proper sequence.
❑ Take precaution against wrong presentation.
❑ Stand on one side of the material presented facing the audience

D. *Follow-up*

❑ Observe reaction of the audience.
❑ Clarify doubts and dispel misunderstanding, if any.
❑ Improve subsequent presentation by deleting irrelevant and old material and adding something new, if required.

6. Qualities of a Good Visual

❑ Easy to use
❑ Suit the audience
❑ Simple and direct
❑ Attractive
❑ Economical
❑ Suit the time and place
❑ In working condition.

7. Limitations

❑ Because of cultural difference, the audience may form a mistaken or distorted impression about the audio-visual aids.
❑ Teaching may be scratchy instead of complete.
❑ Over-reliance over audio-visual aids may convert teaching to showmanship.

Extension Talk

Man can articulate and speak in the form of language. He communicates with others in different situations to fulfill his needs and interests. But all the communication in the form of gossips, chit-chatting and loud talk are not planned. Hence, such talks cannot become extension talk. It is basically different from other talk and discussions. The extension talk can be defined as a "verbal explanation (or presentation or communication) to a group of people, to impart knowledge, by activating the learners".

Extension talk is different from the lecture method. In lecture most of the time flow of information is one way, *i.e.* from communicator to receivers and interaction between the communicator and receivers or audience is very less. On the other hand, the flow of information in extension talk is two way and the interaction between communicator and receiver is much higher as compared to lecture method.

In training programmes, when we are mostly dealing with the adult learners, extension talk is the most suitable and effective method for transfer of know-how. The adults always possess some knowledge and experience, hence they should not be treated as passive listeners. They must get sufficient opportunity to participate so that their valuable experience and knowledge could be utilized. So, extension talk should be preferred over lecture method, while dealing with the adult learners.

Planning of Extension Talk

Planning for an extension talk should be done well in advance. The following elements should be given due consideration, while planning an extension talk.

(i) Audience

The number and type of trainees, their characteristics, background and experience, cultural and social environment, their knowledge of the subject, level of education, their needs and interest etc. would be very useful to the trainer to know, while planning the extension talk and making alterations in the subject matter, whenever needed.

8. Criteria for Selection and Evaluation of Audio-Visual Aids

Selection and evaluation of audio-visual aids may be made on the basis of some items suggested below. The scoring procedure is given on the right hand side.

Items	*Very much* (3)	*Much* (2)	*Not so much* (1)	*Not at all* (0)
The extent to which the audiovisual aid:				
(*i*) Represents a true picture of the topic				
(*ii*) Has the ability to draw the attention of the audience				
(*iii*) Can arouse interest of the audience				
(*iv*) Contributes towards meaningful understanding of the topic				
(*v*) Is likely to bring desirable change attributes?				
(*vi*) Is appropriate to the level of the audience?				
(*vii*) Is worth the time and effort involved?				

"Remember Audio-Visual Aids are Aids only"

(ii) Objectives of Extension Talk

The subject matter and mode of presentation would very much depend on the purpose of the objective of the talk and the end result desired to be achieved by the session. The approach will vary depending whether the objective of talk is to impart knowledge, to change attitudes or to create interest or awareness etc. The objectives of extension talk should be specific, well defined, measurable and achievable in the given time.

(iii) Content

The subject matter should be valid, authentic, factual and applicable. It should be related to the objective and understandable. It should be collected from the various available sources. It should pass through the process of informative reading, selective reading and final reading keeping in view, specifically of the topic, need and interest of the audience and available time.

(iv) Steps

The finally selected content should be divided into some meaningful units or steps so that the learners can understand and retain it.

(v) Order

Different steps or units must be arranged in a logical order or sequence. The order or sequence may be based on the following principles: (1) known to unknown (2) simple to complex (3) concrete to abstract (4) observation to theory and (5) general to the particular.

(vi) Duration

In order to match the final content with available time, the content in each step or unit may be further divided into essential, desirable and possible parts. The essential content in each step is much and should be given by all means. It time permits and need persists, desirable information can also be given and if there is still time at the disposal of the communicator, possible information can also be covered.

(vii) Division of extension talk

For making it effective, an extension talk is divided into four main parts: (1) The introduction (2) The body (3) The conclusion or summary and (4) Questions or recapitulation, which are described herewith.

1. Introduction: The initial part of extension talk is brief opening to capture the audience attention. About 10% of the total time is utilized for motivating the audience. The introduction may be : (1) a humorous anecdote, a success story, a severe challenge, a serious question or anything else to capture the interest of the group (2) Salient features of the topic in brief, without giving details (3) Importance of the topic in statistical and economic terms (4) Brief preview of the session and linkage of the present topic with the previous one.

2. Body: The central theme of extension talk is in its body. Here the message of subject matter is delivered in detail. The content should be according to abilities and needs of the learners. Three-fourths (75%) time is utilized for this part of the talk. While presenting the subject matter, the trainer should try to create a thought-provoking situation and to achieve this purpose, he/she should follow these points:

- ❑ Divide the message into meaningful units/steps, put these units into a systematic order.
- ❑ Identify essential, desirable and possible parts of the subject matter in each important step.
- ❑ Use short sentences and simple language. Simple language has less chance of being misunderstood. It is to be remembered that the trainer's role is to make difficult things simple and not the reverse.
- ❑ Develop specific support information for each main point. Support the major ideas or subheadings with relevant examples, illustrations, anecdotes, outside experiences, reading etc. to make the presentation logical and interesting.

❑ Frame different types of questions to involve learners and to increase their participation.

❑ Prepare/arrange and use variety of visual aids. Example, illustrations and statistics can all be emphasized and made more understandable by the use of visual aids.

3. Summary : If your talk is having several parts, it will be advisable to summarize frequently each part before starting a new part rather than do it at the conclusion of your presentation.

It is always useful to prepare a summary in advance and to put down in the talk outline for reference. It helps the trainer to have better grasp of the talk, if he has once written out the summary for it. It also helps the audience to crystallize the ideas, five per cent of the available time be utilized for this purpose.

Rather than reiterate the major points precisely, in the same words, a slight rewording or rephrasing is always in order. Summary should be brief, precise, logical and easy to remember. The actual words "in conclusion" or "to summarize" give the audience a clue that the speaker is ready to finish.

Finally a word about "thank you" to your audiences for their interest and attention is also in order.

4. Question (recapitulation) : After giving summary and conclusion, the trainer should ask some questions from the audience for recapitulation. The learner should be asked to reply these questions. In this way, the trainer can get the feedback that to what extent he/she has been successful in achieving the objectives, 10% time is utilized for this purpose. These questions should be framed well in advance and written in a talk outline.

A blank proforma developed for planning an extension talk, and an example how to plan an extension talk, are given at Annexure II and III respectively.

Delivery/presentation of Extension Talk

❑ The trainer should be on the time for the talk and check for visual aids and other necessary teaching aids and physical arrangements.

❑ Increase interest/motivation of the audience by giving an effective introduction of your topic, make the objective clear.

❑ Present your subject matter step by step, in a systematic order.

❑ Your presentation will be more effective, if you allow involvement of participants learners.

❑ To focus the attention of the learners on the subject, make effective use of different types of A.V. Aids. A visual presentation captures their eyes as well as their ears.

❑ In some sessions, humour may also be evolved naturally. If it is comfortable for you, encourage informal atmosphere of humour. Humour relaxed the participants and enhances their learning.

❑ Use pauses and silences effectively, to emphasize a point and to encourage learner's reaction.

❑ Use gestures, postures and movement effectively. Maintain eye contact with different segments of the audience, which in turn improve the effectiveness of the talk. Try not to your back. Avoid playing with a chalk, ruler, keys etc. and be well balanced from emotional point of view.

❑ Appear confidence, comfortable and enthusiastic about your role as a trainer and about the content your are presenting.

Types of Questions

To involve your participants or to increase their participation different types of questions can be used as an important tool. The details of the questions with their characteristics are described herewith.

(i) Direct question

This type of question is aimed at one person. For example, "Mr. A. Saxena, please tell me, what are some ways, we can improve sales in this region?" The purpose of asking this question is to draw the attention of a person, who is not attentive in the class.

(ii) General question

This type of question is known as indirect question or overhead question. This types of question is asked to the entire group, so that all the trainees in the class are involved. For example, "A question is put to all the trainees, what are a couple of ways we can increase sales?" It is an indirect question. The purpose of asking this type of question is to involve all the learners to solve a particular problem.

(iii) Reverse question

In this type of question, the trainer returns back the same question to the learner, who has asked the question. This techniques is generally used where the communicator has got the impression that the learner knows the answer of this question and he is asking the question just with the purpose of pulling the leg of communicator/trainer.

(iv) Pass on question

In pass on the question, trainer passes the question, asked by one learner to some other learner from the group. The question is passed on to the person, who can answer this question very well. This type of question is very effective tool to increase the participation and to utilize the group expertise.

Technique of Asking Questions

"How to ask a question", is an art. An expert trainer should be well versed with the art of asking questions. Following points should be kept in mind, while asking questions :

- ❑ Talk clearly and loudly.
- ❑ Choose the words carefully
- ❑ Avoid the use of unfamiliar and ambiguous words
- ❑ Make the question short
- ❑ Make intention clear
- ❑ After asking question, wait for some time
- ❑ Always put questions to passive learners/trainees.

HANDLING GROUP AND INDIVIDUAL SITUATIONS

Group Situations: Probably no group would remain constantly in any one of the following classes, but all groups fall into those categories on occasions:

Situation	Remedy
(A) Bright, Active, responsive group.	1. Be well prepared.
	2. Give it to them fast.
	3. Ask tough questions.
	4. Don't pit yourself against them.
	5. Pit them against each other.
(B) Resistant, antagonistic group (Lack of understanding, experience and interest)	1. If you must face issue frankly and find out one or two who are responsive
	2. Find out the cause and correct it to show sympathy for their situation, it possible.
	3. Work through their personal solution.
	4. Don't break down the stone wall with a bulldozer.
	5. Find a loose stone and break the foundation and brick at a time.
(C) Slow, passive group.	1. Do more telling than usual.
	2. Ask simple but provocative questions.
	3. Thoroughly explain the topic.
	4. Use effective aids to understanding.
	5. Show lots of enthusiasm yourself.
	6. Don't go too fast.

Individual Situations : There are two general classes of individuals. Those who talk and those who don't talk. However, within these broad classification there are some following situations which may cause trouble.

Situation	Remedy
Talks too much:	Cut across his talk with a summarizing statement and direct a question to someone else. Talk to him during a break, thank him for his input, ask him to slow down a bit so that other may participate. If he is difficult get the audience on your side, they take care of him.
Quick, helpful:	This man has the right answers but keeps others out. Cut across him tactful by throwing a question to someone else. Be sure he understands you, appreciate his help. Suggest, "Let's get more opinion" Use him to summarize.
Rambler:	When he stops for breath, cut of by thanking him politely, rephrase one of his statements and pass on. Refer to board and ask which topic he is discussing.
Arguer:	The participant who argues might be placed in a blind Spot "right next to you". Pretend not to hear. Of course recognize legitimate objections and side with him when possible. Talk to him privately and ask for this help. As a final step you may ask him to leave the session/class.
Obstinate:	This is usually someone who does not see the point. Try to get others to help him see the point. If he is the only one, proceed and tell him to see you after the session/class.

Wrong subject (Off the beam)	Direct attention to topic on the board. Ask the individual, if this is relevant to topic or simply say "It is interesting, but would you hold it until later."
Complaints about management	Tell him the problem is how best to operate under present system. Try to get another reliable member to answer him. Tell them that we can't change policies. Don't waste too much time on management complaints.
Has a problem of his own:	Tackle his problem, if it is pertinent. Get group opinion, then question him on your subject or acknowledge worth of his problem and ask that it will be brought up later.
Racial or political problem:	Frankly state what you can or can't discuss. Problems do exist, but out work must be done anyway.
Side conversationist:	Pause, let others listen to the conversation, walk down. By him and thus draw attention to the conversation. Draw him to your discussion by asking for his opinion. This polite, non-threatening intervention will stop conversation.
Poor choice of words or voice	Help him repeat. Recall his ideas in your own words "in other words, you mean....." Protect him from ridicule.

Skill Teaching Method

The fundamental principles of learning is that people learn trough seeing, hearing and doing. Demonstration are based on the premise : "if I hear, If I forget, if I see, I remember, but if I do, I learn."

While theory takes one to the threshold of learning, practical, open the door of learning. A happy synthesis of both is ideal.

Practical relates to the acquisition of manipulative skills and abilities. The most important aspect of practical is that it involves doing of a job to acquire the necessary competency/skill. Just the theory is know-how, practicals are 'do how' part of the learning process.

The learner must learn the 'why' as well as the 'how' of any thing/subject to be learnt. One may run into trouble and he cannot perform unless he knows how it will be done. Therefore, learning by doing is the best principle of learning skill. A skill teaching is the process of learning by doing.

Definition: Skill teaching is defined as "To train the learner, to perform a job, as quickly as possible, under supervision". In the process of skill teaching, extension worker performs a job, step by step. The learner observes the process and listens to the real explanations carefully with intention to gain ability as how to perform a particular skill. Later on, learner repeats the demonstration given by the teacher/trainer.

Elements of Skill Teaching

There are three major elements involved in the process of skill teaching, which are described herewith:

- ❏ Planning for skill teaching
- ❏ Rehearsal for skill teaching
- ❏ Presentation for skill teaching

(a) Planning for Skill Teaching

Planning is a process of thinking, consultation and documentation about a particular skill. To perform any skill successfully and effectively, a good planning is necessary. The following points should be kept in mind, while planning for skill teaching.

(i) *Importance of job for the learner* : For effective planning, it is essential that a communicator should know. What is the importance of a particular skill for the learner. Why should a learner/farmer should perform a particular skill/job?

(ii) *Previous knowledge/skill to perform the job* : The communicator should try to know, How much the learner knows about the skill to be taught so that he can teach the things which are not known to the learner with respect to skill in question?

(iii) *List out tools/equipment and material* : The communicator should be well versed with the make, quality, quantity, price and utility fo new chemicals/tools and equipment to be used, as it is expected that the learner may ask about it, prior to demonstration. He should be well aware about the locally available tools, equipments and materials required for the skill to be taught. List of all the required equipments/material should be prepared.

(iv) *Division into steps* : For conducting the skill session with confidence, it is essential that the whole job be divided into steps, as it assures understanding of learner and also strengthen the remembrance of learner.

(v) *Identification of key points* : Each skill has certain key points. This must be brought into the knowledge of the learner so that a lasting impression in his mind is created. These key points are important points of any skill, which are essential to perform a particular job/skill.

To maintain the quality of uniform spraying of any insecticide on any crop, the height of the nozzle, the speed of the person and the width of the spray should be constant. Key points are the points, which determines the quality of a skill/task which make the execution of task easier and which prevent damage to the crop/machine and harm to the body.

(b) Rehearsal

Rehearsal is must before performing any skill, because it enhances the credibility of the communicator and also makes him confident to transfer any skill to the learner. The communicator should rehears the skill himself, before teaching a skill to the learner.

Following points should be considered, while rehearsing for a particular skill:

❑ Follow the steps of a skill, as it increases the retention.

❑ Every step has some important key points, test must be understood by the communicator properly.

❑ One must perform skill himself, before teaching to the learners.

❑ No assistants/subordinates are required, while rehearsing for a particular skill.

(c) Presentation

Presentation of skill teaching means 'the farmer is before the trainer or extension worker'. There may be two situations:

(i) The extension worker/trainer has gone to a village/farmer's field to teach the skill to the farmer.

(ii) The farmer has felt a need and has come to the office of extension worker/trainer to learn a particular skill, which he is not able to perform.

The transfer of a skill to the learner/farmer, is not an easy task. It requires a high skill of transferring the skill to the learner/farmer. One must be highly skilled in the areas of technical/subject matter, communication and social and psychological aspects.

There are four important steps in present/transfer of skills to the learner/farmer, which are being described herewith:

(*a*) Prepare the learner

(*b*) Show the learner/demonstration of skill

(*c*) Learner's practice/Let the learner try

(*d*) Had over the job.

(*a*) **Prepare the learner:** The modern environment in which we are living now-a-days is a tenseful environment. The farmers/general public is also living in the same environment. A farmer to whom we are going to teach/transfer a new kill, may face a situation where in any one of his family member may be ill. He has to organize a marriage ceremony of his daughter, he may not have received proper electric supply for irrigation and could not arrange other inputs for his field operations. In such situations, being mentally worried, he can't pay full attention to whatever message/skill, we are conveying/teaching to him. It is, therefore, essential that he should be prepared well mentally as well as physically, to receive the message with full attention. Following points should be kept in mind while preparing a learner/farmer to teach a skill:

(*i*) *Put the learner in relax mood:* The learner should be relaxed by asking about his family welfare, his crop conditions or related things so that he is in a relax mood and rapport may be developed for transferring the skill.

(*ii*) *Tell him purpose:* Why have you visited the learner? Clearly state the specific purpose to the learner the today I have come to transfer this skill to you.

(*iii*) *Utilize previous knowledge:* Prior to perform the skill, communicator asks about the previous knowledge/skill of the learner related to the job. Not only, he should ask to narrate verbally, but also to do the job. During action, communicator should watch and note the mistakes committed by learner and guide the learner to improve upon the mistakes at the time of performing the skill.

(*iv*) *Arouse interest:* The interest can be created in learner by explaining the importance of the skill or profit on account of performing the skill or loss involved by not performing the skills. Economics of the new skill may also help in creating interest in the learner.

(*v*) *Put the learner in the right position:* The learner should be put in the right position so that he can observe the communicator while performing the job/skill, properly. The communicator should also draw the attention of the learner to watch each and every aspect of the skill performed by him.

(*b*) **Show the learner/demonstration:** following steps are essential for demonstrating the skill before the learner/farmer:

(*i*) *Explain the basic knowledge:* Prior to demonstration, one should know the basic knowledge involved in doing a job e.g. what material is required to perform the job/skill etc. The quality, quantity/dose and expiry date of the chemicals etc. used during demonstration should be checked properly.

(*ii*) *Demonstrate slowly and systematically:* Now demonstrate the skill slowly and systematically. Following one step after the other. It should be kept in mind that the steps are performed in a sequential manner, till the last step is covered properly. The summing up after two to three steps will enhance the learning during this stage.

(*iii*) *Instruct step by step :* While performing different steps, sufficient explanation should be given, so as to enable the learner/farmer to understand a step properly. While explaining about a particular step, it is essential that the description about the other step be avoided.

(*iv*) *Stress key points:* It means more emphasis is given on how a particular step is to be performed and why it is to be perform? This process will enhance the understanding of the learner regarding a particular key point.

(*v*) *Speak clearly, fully and patiently:* For a good communicator, it is essential that the expressions are clear and accurate. Only then, the message/skill will be learnt by the farmer/person properly and fully. At the time of speaking, the pitch and speed of the voice should be according to the situation.

(*vi*) *Convey vital information at the right speed:* The important information required to transfer a skill, should also be given with the right speed. We should not be very at the time of explaining vital information in a step.

(*c*) **Let the learner try:** A change should be given to the learner to do the skill. A learner/farmer may not be perfect in performing any skill, unless he has himself practiced the skill again and again in the presence of a communicator/trainer. Following steps should be adopted during the stage of learner's practice.

(*i*) *Examine ability:* The ability of the learner could be examined before he actually performs a skill. It is better to examine ability at intermediary stages. The communicator should observe the learner and correct him if he is wrong.

(*ii*) *Let him do and explain:* The learner should be provided with a chance to do the skill and explain about each and every step simultaneously.

(*iii*) *Let him show again:* The practice makes a man perfect. Therefore, to do a job effectively, without errors, the learner should be given a chance to perform it again and again till he is to do the skill perfectly within minimum time.

(*iv*) *Check the key points:* At the time of practice by the learner, communicator should check the key points. Observing the method of explaining, how and why aspects of each and every step being performed by the learner while doing a task do this.

(*v*) *Encourage the learner:* The encouragement is the input, which involves no cost. Therefore, a learner/farmer should be appreciated whenever he performs a step properly/correctly. This will enhance his motivation to learn a skill.

(*vi*) *Hand over the job:* Following steps should be kept in mind while handing over the job to the learner:

(*i*) *Stress responsibility:* The communicator should emphasize that it is the responsibility of the learner/farmer to understand the skill properly. If he fails in doing it, he will be the loser. He should also transfer "the learnt skill" to the fellow farmers, so that they also learn it and get benefit out of improved skill taught by the communicator.

(*ii*) *Encourage questions:* The learner should be requested to ask any questions he wants. This opportunity, if provided, will enable the learner to remove his doubts regarding the skill.

(*iii*) *Ascertain, if all is correct:* After learner practice, the learner should be asked to perform the skill again with a right speed with no mistakes. This is essential to confirm whether the learner has attained perfectness in performing a particular skill or not.

(*iv*) *Introduce possible helper:* At the end of skill teaching session, the address of a nearby person, who is expert in performing that skill, should be given to the learner/farmer. The learner may consult him as and when needed. This is essential to provide solution required while performing a skill in the absence of communicator/trainer.

The Teaching of Skills (Division into steps)

Sr. No	Steps	Key Points	
		How	*Why*
1.	Selection of spot.	Spots should be selected randomly representing whole field.	To know the real nutrient status of the field.
2.	Cleaning of spots.	With the help of spade.	To avoid soil contamination and for taking pure sample.
3.	Digging of pits.	9″ deep "V" shape pit with the help of khurpi/spade.	Because this is the main nutrient supplying zone.
4.	Taking approx.	500 gm. soil with the help of Khurpi/Spade from each pit.	For mixing the soil from all pits.
5.	Mix the soil.	With the help of hands and also break clouds and remove the pebbles/roots etc.	To make uniform soil for knowing actual soil nutrient status.
6.	Take 500 gm. soil.	First divide whole soil into parts and mix two opposite parts. Again divide into four parts and mix two opposite. Repeat this process till we get 500 gm. soil.	To take true representative/sample of field.
7.	Filling in bags.	With the help of hands.	For sending to Soil Testing Lab.
8.	Putting identification.	By putting (*i*) Farmer's name; (*ii*) No. of field; (*iii*) Name of village on a paper attached to bag.	To avoid mixing/misplacing of sample with other soil samples.

Precautions : 1. Khurpi, Tasla, polybag and Spade should be cleaned properly before taking soil sample from the other field.

2. Polybag should be fresh and well cleaned.

Skill Teaching Outline

A. Basic factors

Subject :	Soil Testing
Topic (skill to be developed)	Taking of Soil Sample.
Aim :	To develop skill in taking Soil Sample.
Equipment/materials:	Kurpi, spade, polybag/Cotton bag, Sutali, Tasla.
Teaching Aids :	_____
Place and arrangements:	
Prepared by :	

B. Outline

Duration	Heading	Content	Visual or Equipment
Introduction	Importance	Soil Samples help on knowing the soil status with regard to nutrients level. It enables the farmer to apply appropriate dose of fertilizers and thus the wastage of nutrients is minimized. Simultaneously crop production increases.	
Basic Knowledge		The soil sample should be taken before sowing the next crop. The soil sample should not be taken from heap of FYM/Soil/Stone shadow of stress and pit. The sample should be drawn from tall corners of the plot/field.	
AIM		The soil should be taking 9″ Deep and slice should be taken from all length of the pit. The auger can also be used for taking sample.	

Extension Talk Outline

Time	Headings	Contents	Teaching aids
1 mnt.	Introduction	So many persons have died while spraying chemicals in the fields. The simple precautions, if kept in mind, while spraying chemicals in the field, can help God to save life of the people. It will also avoid damage to the crop.	
1 mnt.	No wounds	If there are wounds on body parts, the chemical may enter into internal system of body and can cause serious reactions and death of the person. Therefore, persons having wounds on body should not spray chemicals.	
1/2 mnt.	Wind direction	The wind direction is an important point to be considered while spraying chemicals. The person should move in the direction of wind to avoid inhalation of chemicals.	
½ mnt.	Use of mask	The person spraying chemical should always use mask. Otherwise there are chances of serious reactions or death of the person.	
2 mnts.	Spray in the afternoon	While spraying chemicals in forenoon times, the presence of dew on the leaves changes the chemical and water ratio in the solution. It causes reduction in the effectiveness of chemical. To avoid this loss the chemicals should always be sprayed in the afternoon time.	
1 mnt.	Avoid double spraying	The double spraying should always be avoided. This practice not only damages the crop but also leaves some part of the crop unsprayed.	
1 mnt.	Spraying in row	For proper spray the chemical should be sprayed in straight lines or row. The zigzag method of spraying should be avoided.	
1 mnt.	Homogenous speed	The speed of the person spraying chemicals should be caliberated and the homogenous speed should be maintained while spraying.	
½ mnt.	Constant height of nozel	The height of the nozel should remain constant for proper and homogeneous spray of chemicals in the field.	

½ mnt.	Summary	Safe, economic and profitable apart of chemicals in possible when sprayed wind direction, mask is used. Sprayed with homogenous spread and constant height of nozel of the sprayer.
1 mnt.	Recapitulation	1. In what direction of the wind chemical should be sprayed?
		2. Why one should use mask while spraying chemicals?
		3. What is the importance of homogenous speed and constant height of no. of sprayer?

CCS Haryana Agricultural University Extension Education Institute Nilokheri (Karnal).

Plan of Extension Talk

Topic:	Precautions while spraying chemicals in the field.
Duration:	10 minutes.
No. of participants:	25
Type of participants :	SMS/SDAOs/MTs/Scientists.
Participants' knowledge:	Participants have some knowledge of the precaution while spraying chemicals.
Aim:	To improve the knowledge of the SMSs/SDAOs/M.Ts/ Scientists regarding precautions while spraying chemicals in fields.
Source material for talk:	1. Handbook of Agriculture.
	2. Package of practices Kharif Crops/Rabi Crops. CCS H.A.U., Hisar.
Main Teaching Aids:	1. Full suspense charts.
	2. Flannel Strips.
	3. Blackboard.
Material for participants	1. Improve your Communication skill.
Teaching place and its arrangements	Well equipped Conference Hall,
	Extension Education Institute, Nilokheri.

Chapter 4

TRAINING

1. Training is purposeful intervention in bringing about desired result. The result may be visualized in terms of commonly known behavioural dimension such as increased *awareness, understanding, knowledge, skill and formation of positive attitude.*

2. Moreover, the desired outcome can result because of both learning and unlearning through training. While learning can enhance and enlarge behavioural dimension, unlearning through training on some of the aspects misunderstood can equally sharpen the learned behaviour. Therefore, training as intervention is viewed in a broader perspective and not in a restricted sense of imparting information about a particular aspect. Such intervention can be seen along with most generalized and fundamental purpose served through training.

3. It is possible to look at the basic purpose of training in the following diagram.

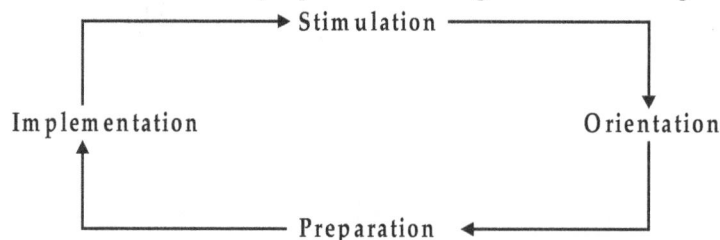

4. The cycle suggests that any training programme may serve the purpose of stimulating people towards improved performance on their job : orienting them about the possibility of acquiring knowledge and competencies, equipping them with needed competencies and finally preparing them to put the acquired, knowledge and competencies into practice. All the above stated aspects may not follow a sequence and necessarily all of these may and occur at one time in training programme.

This model has specific relevance to entrepreneurship training programme.

Entrepreneurial Training Programme

1. Training in entrepreneurship provides an impetus to the potential and budding entrepreneurs to acquire a new identity about himself.

2. This is perceived as an approach towards transforming people which serve the purpose of making people aware about their own identity, helping them accept the new identify and finally establishing such identify for entrepreneurial pursuit.

3. In order to take up such task of transformation we can find entrepreneurship training serving the purpose of stimulation, orientation, preparation and implementation in a sequential order.

4. In order to apply the model, one requires an understanding about approaches to entrepreneurship training programme.

Recent Approaches for Entrepreneurship Training Programme

With the increased emphasis of developing entrepreneurs, the country has witnessed a phenomenon change in its approach to create entrepreneurial awareness and generate entrepreneurship in the society. What appears to be most significant is the realization that no approach can meet all requirements in dealing with diverse population scattered over widely dispersed area of the country. Consequently, different types of entrepreneurship development training programmes took a shape in making it as an effective preposition for industrial development. Niesbud as an apex institution in the country took initiative in emphasizing the need to conduct target oriented EDP's'. For this the institute evolved model syllabi for different target groups which are widely accepted in the country. Recently, the institute added other dimensions to its approach by emphasizing the need to consider location specific and product process oriented EDP's. Thus the recent approaches to entrepreneurship training may be seen as :

New Dimensions

❑ Target Oriented
❑ Location Specific
❑ Product and Process Oriented

Target Oriented

Such EDPs are directed and planned for a specific group which has distinguishing features as compared to other groups. It is because of the fact that target audience vary in terms of their background, experiences, training and exposure to business world. Moreover, the size of the proposed enterprise may also very from one target group to others. These variations demand for the matching training inputs with varying degree of intensity to bring these entrepreneurs to the threshold from where they can launch their enterprise and be able to manage them successfully. For example, a group of MBA entrepreneurs may require less intensity on management skill development input as compared to other aspects. Similarly, for a group of entrepreneurs, who want to set-up considerably larger units, may require high degree of management skills rather than those who are going to set-up considerably smaller units. For this purpose, various target groups have been identified. For example, the identified and operational target groups are :

(*i*) General Entrepreneurs

(*ii*) Science and Technology Entrepreneurs

(*iii*) Women Entrepreneurs

(*iv*) Educated Unemployed Entrepreneurs (SEEUY)

(*v*) Ex-servicemen Entrepreneurs

(*vi*) Rural Entrepreneurs including weaker-sections

(*vii*) Artisans Entrepreneurs

(*viii*) Tribal Entrepreneurs

(*ix*) Physically Handicapped Entrepreneurs

Location Specific

In order to reduce regional imbalances and the imbalances that may exist between progressive and non-progressive areas, it is of considerable importance to plan EDPs depending on the characteristics of the area itself. The Govt. of India has attached significance to this aspect by categorizing of the districts in the country according to different level of backwardness. Out of 432 districts Government has divided backward areas in the country in three categories, *i.e.*, 131 districts in category 'C'. Towards, this Government has provided several incentives to enable entrepreneurs to establish industrial undertaking in backward areas. These included concessional finance extended by the all India Term Lending Institutions, out right subsidy on fixed capital investment, preferential treatments in the grant of industrial licenses, etc. Even the programme in rural and urban area has to be viewed differently. Thus location in terms of backwardness, urbanization, concentration of specific target group in a particular area, etc. may be considered under location specific EDPs.

Product and Process Oriented EDPs

The late eighties is experiencing a shift from Target Oriented to product and process oriented EDP'. Such EDPs are organised for a group of prospective entrepreneurs who opt for enterprises having set product line or process such as plastics or electronics or construction materials or food technology etc. It seems to have a great future. But some of the efforts made recently were found suffering from the lack of appreciation of the role played by inputs, other than product and process orientation. In the absence of such comprehensive programme, the acquired knowledge and skills about product and process will not automatically result in development of entrepreneurial quality, competence for enterprise launching and ability to manage which are crucial to start and ensure success.

Entrepreneurial Approach in Training

The entrepreneurship training differs with other type of training in terms of its nature, scope, result, target group, post training activities etc. Promoter of training to the small business requires to undertake a business like approach. This business like approach to training must also, however, be entrepreneurial in delivery and marketing. An wholistic view of entrepreneurial approach in training in contrast with traditional approach can be depicted in the following exhibit.

Entrepreneurial & Conventional Training Approach

Conventional Approach	*Entrepreneurial Approach*
Major trainer focus on content	Major focus on process of delivery
Led and dominated by trainer	Ownership of learning by participant
Training expert hands down knowledge	Trainer as fellows learner/facilitator
Emphasis upon 'know that'	Emphasis upon 'know-how' and know-who
Participants passively receiving knowledge	Participants generating knowledge
Sessions heavily programmed	Sessions flexible and responsive to needs
Learning objectives imposed	Learning objectives negotiated
Mistakes looked down upon	Mistakes to be learned from
Emphasis upon theory	Emphasis upon practice
Subject/functional focus	Problems/multi-disciplinary

Training can be of immense use in visualizing and emulating the training process in entrepreneurship.

Training Process in Entrepreneurship Development

The emphasis of entrepreneurship training is on transformation of people from 'general' to 'specific' i.e., 'person' to entrepreneur'. Whenever it moves during training process, the same starting point become the focal point at the end with a difference. The difference in terms of a new identity of a person as an entrepreneur is the measure of the effective of training.

The process as a whole comprises of three partners interlinked at three different stages. This can be presented the following diagram :

Training Process in Entrepreneurship

The model is useful to visualize the entrepreneurship training programme as a whole and also each event and series of event which make up the programme. The three partners are: The participants, the Training institutions and the Sponsoring Organisation against three stages, i.e., pre-training, training and follow-up. The model depicts that the participants, the sponsoring organisation and training institution becomes the antecedent to entrepreneurship training and the improved behaviour on the part of participants resulting in setting up of enterprise becomes the dependent variable. Based on the feedback from the participants and the sponsoring organisation, the modification is made in training programme which can lead toward improved training and greater success of organisational objective.

In this process the training continues to bring improvement at individual level, thus changing identity and helping the organisation to achieve the training the objective.

All the three phases occur with participant, the sponsoring organisation and the training institution.

The Learning for the Participants

A. Pre-training

Perceived need of training and psychological preparedness for the same by the participants are very crucial role in learning through training. The arousal of need in case of persons seeking entrepreneurial career can result because of dissatisfaction with present, higher aspiration, higher innovativeness, searching for independence, etc. People experiencing such needs develop an expectation to search alternative, and economic opportunities to generate economic activities. The participants point of view and motivation will determine his focus of attentions towards learning. An answer to the following question may determine the mental preparedness of the participant to get entry into the training.

❏ Does it suitable for him?

❏ Dies he feel qualified for it?

❑ Does he possess necessary background?

❑ Is it suitably timed?

❑ Does he want to attend because there is no alternative?

❑ Does he want because he is idle?

❑ Does he want to attend for the satisfaction of somebody else?

These questions can be equally useful for the trainer to create a situation, before the participants where they themselves develop interest in seeking answers to the questions.

B. Training

Participant brings with him all his expectations in the training programme. He continues to pass through a series of learning opportunities. He starts exploring and assessing his own identify by acquiring knowledge, competencies and attitude towards his set goal. The formation of new identity is established by internalizing what he has learnt and how he is going to translate the learned behaviour into actual practice. Thus the exploration and experimentation of identity formation continues during the training. Some questions like: Can I become an entrepreneur? What do I possess and what I Have to acquire? How soon can I acquire? etc. are helpful to a participant during the training to form a new identity as in combination, the participant them focuses his attention what seems to him useful, stimulation engaging and in line with others.

C. Follow-up

The participants leaves for his destination after the training is over, if the finds training useful and relevant he experiences a new pattern of behaviour, as some what changed person. With his motivation heightened and new enthusiasm from the satisfaction of learning he is eager to use in real life situation what he has learned. In case of entrepreneurs he starts the process of launching the enterprise. This is the stage when he starts questioning the usefulness of items learned through training. This is the stage of action which depends primarily as continued learning. It is possible that he may need to add or delete something in his learning basket by experiencing in actual field situation, something practicable, useful, relevant, applicable are retained and reinforced for further use. Therefore, the follow-up the participants is primarily related looking back to his learning and making use of the learning for the set goal for himself and a goal set for him by any organisation or individual.

Training Process for the Sponsoring Organisation

In the entrepreneurship training programme, the participant's organisations are mostly the organisations sponsoring the programme. In general entrepreneurship training is being organised by training institutions, promotional organisations and voluntary agencies. Since the programme is organised without charging any training fees from the participants, the financial implication of programme becomes a vital issues. The sponsoring organisation participate by way of providing financial assistance for organising the training programme. Some of the prominent sponsoring organisations are: Department, of Science & Technology (DST), IDBI, IFCL, Lead Bank, Directorate fo Industries of different States, small Industry Development Organisations etc. The role of such organisations can be seen right from policy making to policy implementations with the help of existing resources in the country. These organisations are not directly concerned in organising programmes but they do support the programme by way of financial assistance, helping in designing, providing necessary guideline in conducting and supporting the programme through systematic follow-up. All the activities can be divided into per-training, training and follow-up phases.

A. Pre-Training

The sponsoring organisations undertake a number of activities before the training programme commences. These activities are :

1. Deciding the policy related to entrepreneurship development through training.
2. Formulating overall objectives of sponsoring training programme.
3. Deciding promotion of entrepreneurship in terms of the target group and location for the programme.
4. Identifying the training institutions.
5. Deciding about the norms of financial assistance.
6. Inviting proposals for the programme.
7. Scrutinizing proposals.
8. According sanction to the proposal etc.

Availability of finance for training programmes with the sponsoring organisations does not guarantee in any way that the objective set for the entrepreneurship programme will be met easily. Sometimes it has been observed that the sponsoring organisations show their willingness to extend financial assistance to the institutions as the training institution themselves are not in position to take up the task because of their own priorities. It may be mentioned here that there are limited specialised training institutions and promotional organisations do vary with their priorities and organise the training programme according to their own priorities. However, a systematic preparation as pre-training activities helps in achieving the target set by the sponsoring organisation.

B. Training

During the training phase, the role of sponsoring organisation appears to be quite limited. Besides making the finance available for the training programme in time monitoring the activities of the programme and extending emotional and moral support to the training organisation becomes crucial.

The sponsoring organisation expects that the programme will be conducted in accordance with objectives and spirit of developmental approach adopted in training institution will remain unchanged. It has been experienced that the organisation did help the training institution in arranging in-plant visit of the participants by making use of constants with existing entrepreneurs trained earlier under the sponsorship of the organisation. This kind of help rendered by the Department of Science & Technology in some of the programme have been quite useful.

C. Follow-up

The involvement of training institution is not limited to organising the programme only. It extend to provide follow-up support to the trained candidates also. The responsibilities of sponsoring organisation is considerably increased at this stage. The progress is monitored both by the training institutions and the sponsoring organisation. But follow-up record on a long term basis is maintained by sponsoring organisation like DST directly contacting the trained entrepreneurs and monitoring their progress in setting up the enterprise. It is not only the question of simply asking the progress by the trained candidate but also rendering help is the participant who finds some problem in his endeavour. In some cases even the training organisation indicate the perceived problems by the entrepreneurs and request the sponsoring organisation for definite action and help. Whenever it is attended properly, by the sponsoring organisation, in general immense interest and confidence on the part of the trained participants. There are occasion, when the participants have expressed their problems directly to organising entrepreneurs meet, displaying entrepreneurs products, providing place in industrial fair for entrepreneurs product display add to the follow-up activities by such organisation.

Training Institution

Training institution is expected to act as a link between the sponsoring organisation and the participants in fulfillment of objectives laid down by the organisation. The role and responsibilities of training institutions can be seen in terms of implementation of ideas, policies and programmes. The entrepreneurship training, in particular is mainly geared and performed by the training institutions. Therefore, as series of activities forms the part of training institutions which may be divided into three phases :

Pre-Training

This can also be termed as "preparation" phase for the training programme. The process starts with basic understanding of situation that calls for transforming people to be an entrepreneur. An analysis of situation can provide conviction towards the goal achievement. Answer to questions like "why entrepreneurship training? Does it relate to national objective? This kind of conviction is necessary because of the developmental nature of the programme that requires proper planning commitment, and clarity on the part of training institutions. Assessing such commitment and clarity is difficulty proposition but requires an urgent attention to provide a real meaning to entrepreneurship training. Besides this, the training institutions requires to undertake a series of activities to prepare itself for the training programme. Some of the activities are listed below:

(a) Documentation of Entrepreneurial Opportunities in the Area

Before launching the entrepreneurship development training it is considered necessary to document the industrial opportunities in the area. It is expected that in selected district some techno-economic survey indicating availability of raw-material, human, infrastructure facilities, potential demand of goods etc. has already been conducted. Similarly, banks might have prepared the District Credit Plan/Bank report which may further substantiate the potentiality of developing a particular area. The training institutions therefore, need to collaborate with local bank and DIC to prepare project profile based on such documents. These documents may be made available to the participants during the programme with an objective to help them identify a suitable project for themselves.

Entrepreneurship Development

After independence in 1947 India embarked on a system of economic planning with higher priority on industrial development especially in Second and Third Five-year Plan (1956-66). The industrial entrepreneurial function has been facilitated by investment in social overhead capital and other programmes, such as industrial estates. Government has emerged as a major entrepreneur accounting for almost one-half of industrial investment in recent years. Although in some activities government investment has deterred private investment, in other instances government investment has deterred private investment, in other instances government has stimulated demand for industry by investing so as to alleviate bottlenecks and by creating new linkages with other industries. In the decade after independence, public investment, through linkage effects, stimulated the growth in private investment. For example, the 72% increase in private investment, from the First to Second Five-year Plan (1951-61) was associated with a 103% increase in public investment, from the First Plan to the Second one.

Even after the industrial policy resolution of 1956, which extended the number of basic and strategic industries in the public sector, the size of the public sector was not large. Government enterprises accounted for 4% of net-national product (in 1960-61), 4% of the total working force (in 1957-58), and 54% of the net investment in Second Plan (1956-61). All these years, the government has been heavily involved in entrepreneurial endeavours in such activities as the creation of institutions for rural community development, credit organisations for firms of various sizes and types; export promotional schemes; and new organisations to develop transportation and communication, education, health source, and scientific and technological research.

Till 1955, almost 90% of the enterprises in India were small enterprises. About two-thirds of all entrepreneurs were included in the category of enterprises "employing less than ten persons with power, or less than 20 persons without power, using mainly household labour. A 1962 study of urban households throughout India indicated that 27.3% of all urban household heads were self-employed persons in non-farming activities-although 80% of these self-employed persons have no employees. Over 40% of the self-employed person were associated with business a net worth less than 200 rupees, which in most cases in only enough to survive. Most of these persons were in business only because they were unable to find other forms of employment. Starting a petty business with 100 rupees or so may be the easiest occupation to enter. The major sources were reinvested earnings.

This situation let the Government of India to make concentrated efforts for the development of indigenous enterprise.

The Ministry of Industrial Development's Scheme of Self-Employment for Engineers and Educated Unemployed

In order to fight the problem of unemployment among the educated persons, the Ministry of Industrial Development formulated the following schemes :

1. Scheme for Educated Unemployed. In 1971-72 the scheme intended to provide package assistance to prospective entrepreneurs which included in-plant training and also creation of a separate cell at the State level to look after the programme. For this Rs. 35,00,000 was sanctioned to each State with the following break-up.

i.	For training 200 persons	6.50 lakhs
ii.	For setting up an industrial Estate	22.00 lakhs
iii.	Seed money for purchase of machinery	3.00 lakhs
iv.	Establishment charges	2.00 lakhs
v.	Other miscellaneous expenses	2.00 lakhs
	Total	35.00 lakhs

Now govt, has started to give educated unemployment stipend U.P. Government also do the same out of the Rs. 35 lakhs sanctioned to the State Government one third will be in the form of grants and two thirds as loans.

2. Assistance to Young Engineers: The scheme had a provision of rupees three crores and was operated by 29 institutions since 1970. The scheme provided orientation training for three months. The total number of engineers trained by all the above mentioned institutions came to about 3618 by January 1973. Out of this 202 engineers were successful in setting up their own units; 79 secured service after receiving training and 72 were likely to start their projects soon.

In September 1972, the States Ministers of Industries reviewed the progress and finally recommended the following steps to attain the objective of self-employment to Engineers and educated unemployed youth.

(i) *Creation of Separate Cell at the State Level:* The absence of a nodal authority at State level to organize; monitor and guide these schemes was felt. The State Governments were requested to appoint responsive officer to hand the cell, preferably an officer of the Bank of Additional Director of Industries.

(ii) *Institutional Training Arrangements:* After orientation training they most be put to in-plant training in projects similar to select projects.

(iii) *Industrial Estate/Areas:* Separate Industrial estate may be earmarked for the projects to be set up exclusively by the young unemployed engineers with necessary provision for infrastructure facilities and linked up with training programme so that trained personnel may not have to waste their time in search of necessary factory accommodation.

(iv) *Finance:* What projects have been submitted, they should be scrutinized by a committee consisting of representative from SISI, Director of Industries, Bank and other financial institutions both from economic feasibility and technical viability angles. Once the project is accepted by such a committee the bank should not hesitate to advance loans in such cases.

(v) *Consultancy Service:* Financial institutions feel secured if some consultancy service is ready to guide the entrepreneurs in running the enterprise and hence such consultancy service may be instituted.

(vi) *Reservation of Scarce Raw Materials:* A separate quota may be earmarked to the states to meet the requirements of projects set up by persons under these schemes.

(vii) *Reservation of Industries :* Industries like electronics, components, pharmaceutical, chemical intermediaries, etc. are to be reserved for development by the unemployed engineers.

(viii) For setting up ancillary industries to large scale industries and public sector projects, the young engineers should be preferred.

Achievement Development Model

McClelland had opportunity to try in Bombay the achievement motivation development courses in 1963. He set in motion a more ambitious programme in the SIET Institute in the same year – which continued upto 1965. McClelland's main emphasis was on a training course which included the achievement syndrome, Self-study, goal setting and interpersonal supports. (Those interested may refer to 'Motivating Economic Achievement", McClelland, David C. Winter, David G., and others, the Free Press, New York, 1969).

"36 Achievement Motivation Training is especially likely to change men who are in charge of their business probably because they have the scope and independence to carry out new ideas and plans. Furthermore, if a man already is somewhat dissatisfied with himself, but sees himself as someone who can initiate specific action to solve specific problems, he is likely to respond to the training with specific and visible activity. Beyond these findings, there appear to be no important difference between those who changed and those who remained inactive. The changers did not have more money. They were not more Western, nor were they less committed to Hindu beliefs and practices."

SIET Model

Fortunately is SIET, members from various disciplines are working for entrepreneurial development. In each discipline initial thinking was that it had the answer to entrepreneurial development. The development group being economists believed that potential surveys industrial profiles and easy accessibility of resources would so the trick. The management group was certain that a person without management techniques and experiences cannot succeed in his business. Later realisation of the necessity of technical and commercial information for the guidance of the entrepreneurs, led to the establishment of a Documentation Centre. When all these groups joined hand, they realized that their combined. Impact would be far greater than the sum total of their individual impacts. SIET in this process learnt the importance of growth, centres, technology, marketing network, institution building and coordination for mobilizing resource and energies. Thus the integrated entreprenurial development plan emerged as a multidisciplinary long-range plan executed through a well knit institutional frame-work.

The SIET Integrated Model for entrepreneurial development has the following 5 main elements:

1. Local agency to initiate and support the potential entrepreneurs till break-even stage;

2. Inter-disciplinary approach;

3. Strong information support;

4. Training as an important input for entrepreneurial development monitoring and evaluation; and

5. Institutional Financing

The model is designed on the belief that unless the programme is operated through local agency with inter-disciplinate nature, if cannot have local involvement, commitment and full utilization of all potential employment avenues. It must operate through all the agencies in the locality which can help entrepreneurial development. The locality team may, therefore, consist of personnel belonging to agriculture, daily, poultry development, public works, industries and commerce department and both local and lead banks. These wherever there is need, the candidates were sent for in-plant training. During the period, candidates were given stipend.

As post training follow-up, the team helped the candidates to arrange necessary inputs for their enterprise, like finance land and buildings, machinery and equipment raw material, power and skilled persons etc.

The lack of market for many products in backward region, of Kashmir was an unique feature of the State. Hence, the team assumed the responsibility of helping the entrepreneurs in getting the market as well.

The results of this integrated model action were beyond expectation. Within a year 575 units came into being and began functioning in one of the most underdeveloped states of India; giving the clear signal that if one can develop entrepreneurship in Jammu & Kashmir, it can be developed in all the other states.

Karnataka Experiment

The Karnataka Government launched a massive Self-Employment Programme in April, 1974. At the meeting of the SIET Institute and the Department of Industries and Commerce, held during the 3rd week of April, it was decided that SIET Institute should assist the Department of Industries and Commerce in the development of entrepreneurship.

It was initially decided that about 8,000 persons should be trained in Self-Employment. This figure, being large, later was reduced to 4,000.

The might be accomplished, it was felt, by pursuing the following:

(a) acquainting the entrepreneurs in the various tools and techniques required for weighing alternatives and conflicting claims which arise when starting a new enterprise;

(b) overcoming their hesitation in starting the enterprise; and

(c) developing an entrepreneurial spirit with a will to overcome hesitation to take moderate risks in competing with others.

A team consisting of the Joint Director, Self Employment Programme, the Principal Director, SIET Institute and a Faculty Member of the SIET Institute started exploration of the various entrepreneurial opportunities. The study in and around Bangalore revealed several opportunities for the development of ancillaries. It was decided that, initially, only undertaking under the control of the Karnataka Government should be selected for ancillary development. Hence Karnataka Electricity Board, Karnataka State Road Transport Corporation, Mysore Iron and Steel Works, etc. were selected.

The above efforts yielded opportunities for:

1. Karnataka State Electricity Board   25 ancillaries
2. Karnataka State Road Transport Corporation   45 ancillaries
3. Mysore Iron and Steel Works Ltd.   30 ancillaries

With the average intake (purchase) of each unit ranging in value form Rs 1,00,000 to Rs. 2,00,000 or equal to one shift production by the parent organisation.

The Karnataka State Finance Corporation has agreed to finance the projects and the Karnataka State Small Industries Corporation has agreed to allot sheds to the selected entrepreneurs.

Methods of Selection of Entrepreneurs

Advertisements were placed in the papers to attract the entrepreneurs. A team consisting of one Faculty Member and two research scholars administered preliminary and final tests before selecting entrepreneurs for ancillaries.

The selected entrepreneurs were offered pre-selected components or items for manufacture.

It could be seen from the above table that the responses is encouraging if the time lag between selection and starting of programme is minimum.

Another idea was to develop functional industrial estates around industries like bicycle manufacturing, making of Kannada typewriters, etc. Each estate was to comprise 20 to 30 small sub-unite manufacturing components with a other unit assembling the final products.

It was also felt that for such products cooperatives should be encouraged.

Hindustan Machine Tools were requested to give a HMT lathe to be manufactured under this programme. HMT organisation agreed to this proposal and give one lathe for manufacturing. They also agreed to sell all the manufactured by this functional industrial estate. Similar explorations were made with HAL, HMT Watch Factory etc.

The third step was to identify industrial and business opportunities at various centuries and district headquarters in Karnataka. Total support and participation of various industries officials from Karnataka was then ensured. These officials were given 15 days training at SIET Institute in June, 1974 in which they were encouraged to prepare an action plan for their districts, with goals and sub-goals that were to be accomplished in their respective areas.

This action plan aimed at the providing basic guidelines for the industries officers. These industries officers were expected to conduct a survey of their district, identify the entrepreneur and guise them in the selection of enterprise.

Soon after the training, these officers were given material such as investment guides required for various industries, projects available with various institutions and product profiles on some industries and businesses.

Assam Experiment

A more comprehensive trial to SIET integrated model was planned in the State of Assam. On the initiative of the Chief Minister himself the Planning Board Government of Assam, initiated the self-employment programme for the educated unemployed youth in May 1974.

It was decided to open one Entrepreneurship Motivation, Training Centre in each district with trained officers from different departments.

Under the overall framework of planning board, one entrepreneurship motivation training centre was proposed to establish in each district. The work of these centres was to be coordinate by Secretary, Planning Board and Director, Manpower Development at State level. It was decided to establish one Entrepreneurship Motivation Training Centre in each district under the overall charge of Secretary. Planning Board and Director, Manpower Development at State level and a group of officers at district level. The team officers in each EMTC was to be trained by SIET Faculty in the methods and techniques of promoting self-employment programme.

Accordingly, a group of 26 officers drawn from various departments like Agriculture, Veterinary Industry, Employ and Craftsman Training, Public Works Department and Assam Civil Service (I and II), received the trainers training in entrepreneurship motivation development. The 21 days training concentrated the areas like:

1. Socio-psychological aspects of Development of Entrepreneurship.
2. Economic aspects of self-employment and
3. Management aspects of small enterprises.

Towards the end of training, the participating officers draw a 19-point action plan to implement the programme of entrepreneurship motivation development.

To begin with six EMT Centres were established at Mangaldai, Kikrajhar, Jorhat Dhemaji, Dephu and Silchar by the end of October 1973. The senior post officer in the team is called as Senior Special Officer and has the responsibility to coordinate the work of the team. According to the Action plan, the team make intensive effort to draw the attention of unemployed educated youth and select trainees through written test and personal interview as per design discussed in their training. A preliminary motivation training for 15 days is given to a batch of 40 thus selected prospective entrepreneurs with an objective to develop entrepreneurship and strengthen the motive for taking self-employment by treating socio-psychological aspects of entrepreneurship development; economic aspects of self-employment, and management aspects of small enterprises.

After preliminary training, the prospective entrepreneurs are helped in selecting enterprises for themselves. This followed by market survey and preparations of project. In case the enterprise training is needed, the EMTC arranges such training and provide stipend to the candidate during this period. Once, the Project is ready EMPT helps the candidate to arrange necessary inputs like finance, land, building machinery and equipment, raw materials, power and skilled persons etc. The responsibility of ENTC does not end here, but they go further in helping the entrepreneurs in establishing their unit, marketing their goods-services and finally repaying the credit and if needed even in the matter of further expansion of the enterprise.

The Assam experiment has in-built programme of evaluation towards the end of every year by SIET with a view to constantly improve upon past deeds.

The achievement of these six EMTCs during November 1973 to July 1974 has in many respects exceeded the achievement through many other programmes going on even in some developed part of the country.

The result of those two experiments suggests the great potentiality of SIET integrated model to develop entrepreneurship even in backward areas of Indian Union.

Progress made by EMT Centres By the end of March 1975

EMT Centre	No of persons trained	No. of project prepared	No. of enterprise training arranged	No. of proposals sanctioned by bank	No. of enterprises established
Mangaldai	385	235	15	60	60*
Kokrajhar	226	118	6	23	22
Jorhat	254	234	193	71	71
Dhemaji	165	96	20	13	6
Diphu	103	80	32	15	9
Silchar	417	290	305	128	100
	1350	1052	581	310	278

*Some enterprises were established without bank finance.

We have seen that the "intensive campaigns" are not suited to development of business enterprises. The Gujarat Programme is successful but its coverage is selective and small. The SIET strategy as implemented in Jammu and Kashmir and in Assam and the Andhra Pradesh Governments' Crash Programme indicates the road we have to take if the self-employment programmes have to make any meaningful contribution to meet the unemployment situation and to accelerate economic growth. Suspicion of now approach may prove a damper in duplicating Andhra Programme. Moreover, the Andhra Pradesh Programme involves an attitudinal change in the administrative set up which may be difficult to be duplicated in other states without some catalytic agent. This introduction of change in the administrative culture and attitude is the strength of SIRT Programme. It further emphasizes and sets in motion an integrated entrepreneurial development programme. The SIET approach which starts by over coming the apathy of the local officers seem to be a programme that can meet the need of the hour (Malgavkar).

REFERENCES

1. Akhouri MP, Indian Experience of Entrepreneurship Development. Akhouri, MP, An Integrated Approach to Entrepreneurial Development: Assam Experiment, SEDME, Vol (4) 1975.

2. Annonymous, Assistance to Young Engineer and Educated Unemployed, Background Paper, Development Comissioner, SSI GOL, Jan, 1973.

3. Bopardkar, A.P. Entrepreneurship Development Programme in Karnataka with particular reference to Ancillary Development, SIET Experiment, SIET, 1975.

4. Cochran, Thomas C.: 'The Entrepreneur in Economic Change' in Entrepreneurship and Economic Development Ed.

5. Hagen, Everett E.: "How Economic Growth Begins" A Theory of Social Change in Entrepreneurship and Economic Development (*op. cit.* Part-I, chapter 6, pp. 123-138) New York, 1971.

   Jaffri A. Thmmos: Characteristics and Role Demands of Entrepreneurship.

6. Kunkel, John H: "Values and Behaviour in Economic Development" (*op. cit* part-I, Chapter-8), N, Y, 1971.

7. Malgavkar, P. D.: 'Entrepreneurial Development–Review and Approach,' SIET Institute, Hyderabad.

8. McClelland, David C. Winter, David C. and other Motivation Economic Achievement, The Free Press, N.Y. 1969.

9. Nanjappa, K.L.: Intensive Campaign-A pragmatic approach for motivating entrepreneurship for starting small scale industries. A paper read in APO Symposium on Development of entrepreneurial Talent at National Productivity Council, 38, Golf Links, New Delhi, 1970.

10. Phiroz B. Medhora, 'Entrepreneurship in India' Political Science Quarterly IXXX (40 Sept. 1965), pp. 558-575.

11. Schumpeter, J.A.: 'The Theory of Economic Development' in Entrepreneurship and Economic Development, *op. cit.* 1971.

12. Sharma, SVS: 'Stimulation of Entrepreneurship in loss Developed Areas-Some Indian Experiences'. A paper presented in Training workshop and study tour on Small Scale Industry in the least developed countries of Asia, for East and Middle East, sponsored by UNIDO. April 1974.

13. Young Frank, W. 'A macro-sociological Interpretation of entrepreneurship and Economic Development, ed. Peter Kilby, The Free Press, New York, 1971.

Chapter 5

MANAGEMENT OF TRAINING PROGRAMME

Training

Venkatachalan **(1983)** observed that organizations operate in an environment of Fast-Changing Technological Developments. To maintain organizational effectiveness in such situation, the skills of the personnel have to be continuously updated, through training and development.

Chowdhry **(1986)** remarked that training is a process, which enable the trainees to achieve the goals and objectives of his/her organization.

Singh R.P. **(1995)** defined training as the process of changing attitudes, improving knowledge and developing skills of the persons/employees of an organization, so as to enable them, to perform their jobs effectively.

Management of Training Programme/MTP can be defined as the progress of developing efficient and skilled man power of an organization, by effectively coordinating all its essential components in a most congenial organizational climate.

Management of Training Programme

A Process of :

❑ Developing efficient and skilled man-power of an organization by coordinating. Essential components line :

  ◆ Training need Assessment
  ◆ Training objectives
  ◆ Course planning and preparation
  ◆ Course content
  ◆ Course design
  ◆ Training methods
  ◆ Team building
  ◆ In a congenial organizational climate

TRAINING NEED ASSESSMENT

Training Need

Johnson **(1967)** Described training need as a gap between the present level of performance and the standard level of performance.

Laird (1978) Commented that the training need exists when an individual lacks the knowledge or skill to perform an assigned task satisfactorily.

Singh (1995) Training need is defined as the actual technical requirement of a person/employee to perform his work/job effectively.

Importance of Training Need Assessment

- ❏ Help in formulating appropriate training objectives
- ❏ Help in developing suitable course content
- ❏ Accurate and need based training
- ❏ Gap in knowledge/Skill is filled
- ❏ Efficient and skilled staff
- ❏ Production and income is increased

Methods of Training Need Assessment

- ❏ Interview method
- ❏ Knowledge test method
- ❏ Performance test method
- ❏ Observation method
- ❏ Workshop method
- ❏ Perception by self
- ❏ Perception by supervisor
- ❏ Combination by self and supervisor

Steps for Training Need Assessment

- ❏ Determine ability level
- ❏ Determine job need level
- ❏ Determine training need

Criteria for Determining Training Needs

Level of Ability	Level of need of the job	Type of training need
Low	High	HTN/CTN
Moderate	High	MTN
High	High	LTN/CGTN
Low	Moderate	MTN
Moderate	Moderate	LTN
High	Moderate	LTN
Low	Low	MTN
Moderate	Low	LTN
High	Low	NTN

COURSE PLANNING AND PREPARATION

Per Course Planning

- ❑ Formulating Training objective
- ❑ Course content
- ❑ Course design
- ❑ Training methodology/Training methods
- ❑ Allotment of topics to faculty
- ❑ Preparation of Teaching AIDS/A.V. AIDS
- ❑ Field visits
- ❑ Initiating Financial Approval
- ❑ Letters for nomination of trainees
- ❑ Preparation of course performance
- ❑ Faculty meeting
- ❑ Invitation to guest speakers

Infrastructure for Training

- ❑ Building
  - ◆ Lecture Hall
  - ◆ Hostel
- ❑ Faculty
  - ◆ Teaching faculty
  - ◆ Supporting staff
- ❑ Funds

Physical Facilities

- ❑ Conveyance
- ❑ Recreation
- ❑ Library
- ❑ Stationery
- ❑ Class room accessories
- ❑ A.V. Equipment
- ❑ Duplicating Machines
- ❑ Video cell
- ❑ Boarding and Lodging

TRAINING OBJECTIVES

- ❑ Training objectives are the directing of a training course also called as course objectives.

Types of Training Objectives

- ❑ General
- ❑ Specific

Considerations while Formulating Training Objectives

- ❑ Based on training need
- ❑ Well Defined/Clarified
- ❑ Realistic/Accurate
- ❑ Achievable
- ❑ Measurable
- ❑ Time bound
- ❑ Should be quantitative
- ❑ Should specify change in:-
    - ◆ Attitude
    - ◆ Knowledge
    - ◆ Skill
- ❑ Use action verb
- ❑ Follow criteria of smart

COURSE CONTENT

Qualities

- ❑ Based on training need
- ❑ Based on training objectives
- ❑ Essential/Desirable/Possible
- ❑ Emphasis on improvement in :
    - ✛ Attitude
    - ✛ Knowledge
    - ✛ Skill
- ❑ Divide into :
    - ✛ Theory
    - ✛ Practicals/Tasks
    - ✛ Field visits
    - ✛ Assignments
    - ✛ Library studies
    - ✛ Recall sessions
    - ✛ Video film show
- ❑ Help in achieving :
    - ✛ Organizational goal
    - ✛ Individual goal
- ❑ Improvement in competence

Components of Course Content

- ❑ Theory/Inputs
- ❑ Practicals/Tasks

❏ Field Visits/Study tours
❏ Assignments
❏ Library studies
❏ Video films

Course Design

❏ Cover all course content
❏ Motivating/Interesting
❏ Active involvement of :
    ✛ Trainees
    ✛ Trainers
❏ Combination of training methods
❏ Contrast in activities
❏ Simple to complex
❏ Logical sequence
    ✛ Attitude
    ✛ Knowledge
    ✛ Skill
❏ Divide into modules
❏ Break up :
    ✛ Theory   15%
    ✛ Practical   60%
    ✛ Field Visits/Study tours   14%
    ✛ Assignments   2%
    ✛ Video film show   2%
    ✛ Library studies   2%
    ✛ Recall session   3%
    ✛ Inaugural/Concluding session   2%

Monitoring and Evaluation

Singh (1995) The monitoring of a training programme has been defined as to assess and control the progress of day-to-day activities of a training course for its effective organization.

Criteria for Monitoring a Training Course

❏ Course objectives
❏ Training Methodology/Methods
❏ Course content
❏ Quality of instructions
❏ Hostel facilities
❏ Use of teaching AIDS
❏ Field publicability

❑ Change in
+ Knowledge
+ Skill
+ Attitude

Important Monitoring Techniques

❑ Pre and post-tests
❑ Daily tests
❑ Weekly tests
❑ Recall sessions
❑ Mid reviews
❑ Back home application

Steps for Evaluation

❑ Development of evaluation performance
❑ Getting the proforma filled from trainees
❑ Processing/Analysis of data
❑ Preparation of reports/Results

TRAINING METHODOLOGY/METHODS

Types

❑ General methodolgy
+ Lecture method
+ More theory, less practicals
+ No module method
+ Less involvement of trainees
❑ World Bank/E.E.I. methodology
+ Conceptual explanation
+ Planning
+ Practice/Rehearsal
+ Presentation
+ Appraisal
❑ Coverable methodology
+ Divided in groups
+ A task is given
+ Observation by a coach
+ Review in groups
+ Presentation in general session
+ Input, if needed
+ Planning for improvement

+ Again a task is given
+ This cycle is repeated till perfection

Training Methods

❑ On the job training methods
  + On the job training
  + Job rotation
  + Guidance and counselling
  + Syndicate methods (small group)
❑ Simulation methods
  + Role plays
  + Case method
  + Management games
  + In basket exercise
❑ Knowledge based methods
  + Lecture
  + Extension talk
  + Group discussion
  + Seminar, symposium
  + Brain storming
❑ Skill based method
  + Assignments
  + Practice after demonstration
  + Task
  + Role plays
  + Skill teaching
  + Workshops

Chapter 6

PROJECT

The rapid growth of population has created problems of unemployment and under employment in such countries. An underdeveloped country suffers from a chronic deficiency of capital resources. The capital per capita is very low to the tune of $350. It is the opinion of most demographers that population pressures are like to increase still further in future in the under developed countries. As such it become necessary to step up the rate of development in order to outstrip the rate of population increase.

Therefore, it is better to emphasize the need for comprehensive economic planning for a backward, under developed economy on the ground that it assured a high rate of economic growth through a quicker process of capital formation. Hence, sound and effective planning is necessary for development.

Project is an investment activity in which financial resources are expended to create capital assets that produce benefits over an extended period of time. That's why projects are often referred to as the cutting edge of development. Project preparation is clearly not the only aspect of fisheries development or planning. Identification of national fisheries development objectives, selecting priority areas for investment, designing effective price policies, and mobilizing resources are all critical. Unless, projects are carefully prepared in substantial detail, inefficient or even wasteful expenditure is almost sure to result–a tragic loss in nations short of capital.

Often projects form a clear and distinct portion of a larger, less precisely identified programme. Again, all we can say in general about a project is that it is an activity for which money will be spent in expectation of returns and which logically, seems to lend itself to planning, financing and implementing as a unit. It is a specific activity, with a specific starting point and as a specific ending point, intended to accomplish specific objectives. Hence, project acts as a "time slice".

It will have a well-defined sequence of investment and production activities and a specific group of benefits, that we can identify, quantify and usually in fisheries projects, determine a money value for.

Its development can be pictured as a progression with many dimension-termporal, spatial, social-culture, financial, and economic. Projects can be seen as temporal and spatial units, each with a financial and economic value and a social input that make up the continuum.

Therefore, project is the smallest operational element prepared and implemented as a separate entity in national plan as a part of development.

An investment project may be anything from a single programme to an entire integrated programme that includes the entire following programme:

(a) Fish pond/Aquarium Business

(b) Hatchery

(c) Feed plant

(d) Ice plant

(e) Cold storage

(f) Processing plant

(g) Wholesale and retail market

(h) Training, Extension etc.

Advantages of Projects

1. The project gives us an idea of cost year by year, so that those responsible for providing the necessary resources can do their own planning. Project analysis tell us something about the effects of a proposed investment on the participants in the project, whether they are farmers, small farms, governments enterprises or the society as a whole.

2. Projects enable a better judgment about the administrative and organizational problems that will be encountered.

3. The project encourages conscious and systematic examination of alternatives.

4. Another advantage of the project is that it helps contain the data problem.

Limitations of Projects

❒ The quality of project analysis depends on the quality of the data.

❒ It is impossible to quantify completely the risk of a project.

❒ Project analysis is a species of what economists call partial analysis.

❒ Another limitation of the project is an underlying conceptual problem about valuation based on the price system.

Project Cycle

There is a natural sequence by which projects are planned and carried out and this sequence is called 'Project cycle'. International development agencies tend to use the World Bank methodology [Baum, 1982]. Under this scheme the cycle is broken down to six stages.

1. Identification or conception

2. Formulation or preparation

3. Appraisal or analysis

4. Implementation

5. Monitoring and control

6. Evaluation

I. Conception or Identification of the Projects

It is the first phase of the project cycle and here we find or identify potential or suitable projects. There can be many sources from which ideas may come for the identification of good projects.

(a) Ideas for new projects can evolve from the present programmes.

(b) Analysis of import and export trends may also brings in new ideas.

(*c*) The most common will be well informed technical specialists and local leaders – A survey of the state or district to project the future needs over the next decade or so will also enable to identify potential projects.

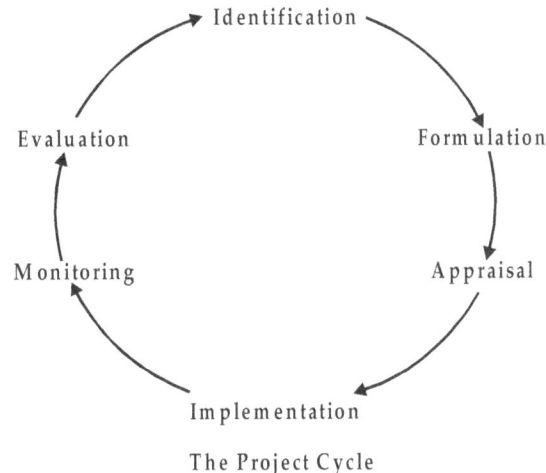

The Project Cycle

(*d*) By investigating local markets.

(*e*) By review of old projects.

(*f*) By observing experience elsewhere.

For a new project these ideas are screened with the assistance of subject matter specialists, experts, engineers Vetynarians/Bankers/Managers economists etc. who have specialized knowledge of factors affecting the feasibility of projects In the various fishery sectors, the screening should be continued until the list for a good project is narrowed down to the most desirable possibilities.

In identification phase, it is also important to see whether the project is implemented in high priority areas and whether on *prima-facie* grounds the project is economically feasible. It is also imperative to identify problems and objectives of the projects and whether the government gives sanction for the project implementation or not.

The important stages in the process of identification are:

(*a*) Preliminary study

(*b*) Pre-feasibility study

(*c*) Feasibility

A pre-feasibility study is listed below:

(*a*) The economy and national status of fisheries.

(*b*) Biological review (resource base, ecology, ocean conditions etc.)

(*c*) Technical review (vessel, gear, infrastructure, posts etc.)

(*d*) Processing, marketing and distribution (including market functionaries, indebtedness to traders, fish transport system)

(*e*) Socio-economic review (human factors, manpower requisite, institutional arrangements)

(*f*) Description of the projects (its status in fisheries sector and impact on the national economy).

(*g*) Status of state fishing and marketing corporations, but also the functioning of cooperatives, banks and other services and an assessment through the ministry of finance and planning of the availability of foreign exchange for fisheries development.

II. Formulation or Preparation

The following points are considered while formulating the projects. The location of the project and project site must be based in technical analysis and technical feasibility of the project. The location of the project depends upon available physical resources, market conditions. Marketing facilities, alternative investment prospects, administrative experience, farmers objectives, technical skills, motivations, demand for products etc. Technical analysis must take into consideration all aspects of technology to be used in the project, and account for all inputs of goods and services. Assessment of suitability and adequacy of natural resources in advance based on the scientific investigations is also essential. Alternatives to the resource use are to be considered in formulation of the project. Due consideration is to be given to all the aspects such as technical, financial, commercial, managerial, organizational, social, economical etc. in the formulation of the projects. Identification of the missing links in the infrastructure system particularly in relation to adequacy of communication systems, markets and storage facilities is important.

Aspect of project preparation and analysis :

According to RIPMAN, 1964, project preparation and analysis can be divided into six aspects:

(a) Technical aspects

(b) Institutional-organizational-managerial aspects

(c) Social aspects

(d) Commercial aspects

(e) Financial aspects

(f) Economic aspects

III. Appraisal or Analysis

Appraisal should take place before the implementation of the project. When a project is fully prepared it is appraised before being accepted as an investment suitable for borrowing. A team of independent experts appointed by government, the project sponsor, the funding agency, or the multilateral bank concerned undertakes appraisal.

The objective of appraisal is to check the thoroughness of the project by making a completely objective and independent study of the project as it has been presented, data have to be checked for reliability, consistency, the reasonableness of its projections, its accuracy in calculations and the validity of its assumptions. It is also necessary to examine the banking, administrative and commercial structures, which will be involved in project implementation and to ensure these have been properly conceived.

There are five criteria for appraisal of fisheries projects.

1. Technical review

2. Commercial review

3. Organization and Management review

4. Financial review

5. Economic review

IV. Implementation

This is the most crucial phase of the project cycle. The secret of successful implementation depends upon the extent of realism put into the plans drawn before hand. It is often not uncommon, to notice our plans getting deviated from the reality. Here the role of prudent decisions by the personnel incharge of implementation to tackle the situation comes into play. Project implementation can be divided into three different periods viz., Investment period, Developmental period and Full-production period.

V. Monitoring

Monitoring is the timely collection and analysis of data on the progress of a project, with the objective of identifying constraints, which impede successful implementation. This is highly desirable, particularly when projects fail, to be completed as per time schedule or in the process of attaining the set goals. It is imperative to get the feedback on the problems faced so that effective measures can be taken up to plug the deficiencies, which hamper the speedy implementation. Monitoring has to be done continuously to offset various shortcomings that crop up from time-to-time with regard to various aspects of implementation.

VI. Evaluation

This is last phase of the project cycle. It is not confined to the completed project. Evaluation can be done several times during the life of a project. In the evaluation process, it is important to see how far the objectives set out in the project are achieved. Deficiencies, snags or failures to achieve the objectives may be analysed and appropriate solutions to such failures answered. Evaluation process is to be completed in three phases. They are mid course evaluation, concurrent evaluation and ex-post evaluation.

Identification of Project Costs and Benefits

In fisheries projects, costs are easier to identify than benefits because the expenditure pattern is easily visualized. The various types of costs involved in the project are:

❐ Project costs: These include the value of the resources in maintaining and operating the projects for *e.g.* physical goods, land labour, debt service, taxes etc.

❐ Associated costs : Costs that are incurred to produce immediate products and services of the projects for use or sale.

❐ Primary costs or direct costs : These include costs incurred in construction, maintenance, and execution of the projects.

❐ Indirect costs or secondary costs: value of goods and services incurred in providing indirect benefits from the projects such as houses, schools, hospitals etc.

❐ Real costs and nominal costs: costs at current market prices are nominal costs, whereas if costs are deflated by general price index, these are termed as real costs.

❐ Social costs: these are technological externalities and technological spill over accrued to the society due to the presence of projects i.e., pollution problems, health hazards, salinity conditions etc.

❐ Replacement costs: Many aquacultural projects require investments that have different lifetimes. A good example is found in the case of water pumping scheme in which the earthworks and pump platforms may be expected to last twenty-five of fifty years but the pumps themselves may have a life of only seven to fifteen years. In preparing the analysis, allowance must be made for the replacement costs.

Next to identifying the costs, the estimation of benefits is imperative to ascertain the impact of the project. Taking into account two situations *i.e.,* 'with' and 'without' the projects generally does this. The difference is the net additional benefit (incremental net benefit) arising out of the project.

Tangible benefits of aqua projects can arise either an increased value of production or from reduced costs.

- ❑ Increased production
- ❑ Quality improvement
- ❑ Change in time of sale
- ❑ Losses avoided

Other Kinds of Tangible Benefits

Transport projects are very important for aquaculture development. This is for not only from cost reduction, but also from time saving and development activities in areas newly accessible to market.

Intangible Costs and Benefits

Almost every aqua project has costs and benefits that are intangible. These are creation of new job opportunities, better nutrition as a result of improved water supply. Such intangible benefits are real and reflect true values. They do not however lend themselves to valuation.

> - ● Stocking
>   - ❑ Introduce warehousing receipt system.
>   - ❑ Private sector stocking to be encouraged.
>   - ❑ Abolish Stocking Limits Permanently
>   - ❑ Strengthen Futures Trading.
>   - ❑ Tendering for delivery for PDS
> - ● Distribution
>   - ❑ TPDS
>   - ❑ With spread of Employment Scheme, PDS can be gradually phased out

> - ● External
>   - ❑ Keep exports and imports free. Only use tariffs as an adjusting instrument.
>   - ❑ Change Commission on Agriculture Costs and Prices (CACP) to Agriculture Tariff Commission.

Chapter 7

PROJECT FORMULATION

Introduction

Once the project have been identified there begins the process of progressively more detailed programme and analysis of project plans. This process includes all the work necessary to bring the project to the point at which a careful review or appraisal can be undertaken and if it is determined a good project, implementation can begin. In the preparation and analysis of projects, consideration will be given to each of the following aspects.

(a) Technical aspects

(b) Institutional-organizational-managerial aspects

(c) Social aspects

(d) Commercial aspects

(e) Financial aspects

(f) Economic aspects

Evaluation of Investment Feasibility and Criteria for Selection of Fisheries Projects

In order to ascertain whether an aquaculture investment project is feasible or not, a cooperative evaluation should first be conducted by both the biologist and economist. Only those species and projects that are suited to the local environment and are biologically feasible for development should be considered. Thereafter, a socioeconomic study can be undertaken. Basically, an economic evaluation includes both the production and marketing functions.

1. The first requirement for any aquaculture investment project in both the public and private sectors is the availability of suitable land and water resources.

2. The species selected for development should be adapted to the local environmental conditions and the stocking materials and suitable feed should be readily available at reasonable cost. The species should also be fast growing and culture technology should be locally available.

3. There should be no legal constraints on development (this is particularly important for private investors).

4. The products of the investment project should have a high market demand with a reasonable price.

5. The investment project should be financially lucrative compared to other investment opportunities for private investors and should also be socio-economically feasible with alternative means of achieving the national objectives for public investment. Private investors usually use profitability as a measure of financial feasibility when assessing

commercial aquaculture projects, and public officials usually consider socioeconomic benefit-cost and/or the social internal rate of return as measures of economic feasibility along with some qualitative judgments.

In order to evaluate the feasibility of an investment protect in aquaculture, one must consider six criteria:

- Resource availability,
- Environmental suitability,
- Biological feasibility,
- Market potential,
- Economic feasibility, and
- Institutional feasibility,

6. It is also important to realize that many variables ought to be considered for each criterion. Each variable can be assessed as favourable, partially favourable, unfavourable, etc. Each ranking can then be scored (or coded) numerically-weighted or unweighted. Next, a general score or code can be assigned to each criterion after evaluation of all the subscores and codes, and the bio-economic feasibility can be determined by weighting the general score or code for each criterion. This procedure can be varied to suit particular projects.

Summary Sheet for Feasibility Evaluation

Criteria	Variables	Rank of Suitability	Score or Code
Resources	Suitable land area	(a) Available for expansion (b) Limited for expansion (c) Not available for expansion	----
		Sub-score or code	-
	Value of suitable land	(a) Low (b) Average (c) High	----
		Sub-Score or code	-
	Water supply of	(a) Adequate year round. Suitable quality (b) Seasonal shortage (c) Not available	----
		Sub-score or code	-
General score or code for resource available			
Environmental suitability	Water temperature	(a) Well suited (b) Suited after temperature is manipulated during certain periods of the year (c) Not suited	----
		Sub-score or code	-
	Salinity	(a) Well suited (b) Suited after salinity manipulation (c) Not suited	----
		Sub-score or code	-

	pH value	(a) Well suited	
		(b) Suited after pH value is manipulated	
		(c) Not suited	----
		Sub-score or code	-
	Tidal flushing	(a) Well suited	
		(b) Suited after tidal manipulation	
		(c) Not suited	----
		Sub-score or code	-

General score or code for resource availability

Biological factor	Breeding	(a) No breeding problem, spawning in captivity, fry available from hatchery.	
		(b) Supply of fry relies on captured wild gravid females, or on the catch from native waters, but availability of fry is not limited at present.	
		(c) Availability of fry from natural waters is limited but the breeding problem is expected to be solved in the near future.	
		(d) Availability of fry from natural waters is limited and the breeding problem is not expected to be solved in the near future	----
		Sub-score or code	-
	Feeding	(a) Nutritional requirements of different age stages are known and appropriate feed (artificial or natural) are available at reasonable cost	
		(b) Nutritional requirements are partially known.	
		(c) Nutritional requirements are not known.	----
		Sub-score or code	-
	Crowding	(a) Adapted to crowding conditions with no major problems in diseases and parasites, and/or with vide range of oxygen tolerance.	
		(b) Some problems with disease and parasites under crowded conditions, or with narrow range of oxygen tolerance	
		(c) Not suited for crowding conditions	----
		Sub-score or code	-
	Growing period	(a) Less than one year	
		(b) One to two years	
		(c) More than two years	----
		Sub-score or code	-

General score or code for resource availability			
Market potential	Price elasticity	(a) Elastic demand with ready market (b) Inelastic demand but elasticity can be improved by market promotion and product development (c) Inelastic demand with limited market	----
		Sub-score or code	-
	Income elasticity	(a) Elastic demand (b) Unitary elasticity (c) Inelastic demand	----
		Sub-score or code	-
	Competition	(a) No competition (b) Competes favorably in price with close substitutes (c) Competes unfavorably with close substitutes at the present but favorably in the future (d) Cannot compete with close substitutes	----
		Sub-score or code	-
	Culture, religion and tradition	(a) Species is currently cultured and preferred by majority of population without socio-cultural limitation (b) Species is accepted by a part of the population due to socio-cultural limitations. (c) Species is not accepted due to religion or tradition	----
		Sub-score or code	-

General score or code for resource availability			
Economic feasibility	Profitability	(a) High rate of return compared with alternatives. (b) Average rate of return compared with alternatives. (c) Low rate of return compared with alternatives.	----
		Sub-score or code	-
	Socio-economic	(a) Average cost per unit of protein, or protein yield per unit of land is favorable compared with alternatives if the national policy concerns animal protein deficiency. Foreign exchange earnings per unit of land or other scarce resources are favorable compared with other alternatives if national policy concerns foreign exchange earnings. Employment per unit of land is favorable compared	

Contd....

		with agriculture activities if national policy concerns employment in rural areas. Combination of all the above-mentioned conditions or any two of them is favorable compared with alternatives.	
		(b) Partially favorable compared with alternatives.	
		(c) Unfavorable compared with alternatives.	----
		Sub-score or code	-

General score or code for institutional criteria

Institutional feasibility	Permit	(a) Easy to get permit (b) Difficult to get permit.	
	Conflicts in use	(a) No conflict (b) Some conflict (c) Strong conflict	
	Use rights	(a) With legal use right (b) Without legal use right	----
		Sub-score or code	-

Reasons for Project Analysis Proving Wrong

When a fisheries project analysis proves to be a poor predictor of the actual outcome of a project it may be that the project design or implementation is at fault, or it may be that the project analyst has done a poor job of incorporating a good project design in an analytical framework.

1. Poor Project Design and Implementation

The most common reason fisheries projects run into problems of implementation may be grouped into five major categories :

❏ Inappropriate technology

❏ Inadequate support system and infrastructure

❏ Failure to appreciate the social environment

❏ Administrative problems include those of the project itself and of the overall administration with in the country.

❏ The policy environment of which the most important aspect is producer price policy

Guidelines for Project Preparation Report

These general guidelines will give an idea about the scope and content of a preparation or appraisal report for an aquaculture or rural development project. Most projects in aquaculture are adaptable to a fairly standard form of presentation. It gives the starting point *i.e.,* the format. It will give the readers of the report a narrative with supporting tables and annexes that gives information and conclusion about the worth of the proposed project without confounding them with unnecessary or extraneous detail.

These guidelines emerge from the combined experience of the food and agricultural organisation, the World Bank and other international lending institutions. The Inter American bank has prepared a comprehensive set of outlines for many different kinds of aquacultural projects. Different elements of a project will need different emphasis depending on the kind of a project.

As far as possible the main text should present the project in a form that a non-specialist can understand specialized backup information-including maps, charts, and detailed tables should be reserved for the annexes or the project file.

The principal elements of project preparation or appraisal report are outlined in the following way. It can include the origin of the project concept in the national development plan in a sector surveyor by a project identification mission. It might mention the government agencies and other organisations involved in the preparation and any external assistance received. It can acknowledge the team that prepared the project and the report and can mention the period in which they work.

Background

A well thought-out and properly constructed background discussion can do much towards establishing the framework of the project and making it intelligible in a broader economic and social perspective. The analyst needs to be very discriminating when choosing material for this part. The only general guidance is that there should be a clear relation between this material and the contents of other sections of the report.

(a) Current Economic Situation

This discussion could mention per capita income dependent on particular moments of imports and exports, balance of payments considerations and the like. It should cover only those features of recent economic developments that have a bearing on the project and on studies of the possible alternatives to the project.

(b) Status of the Fisheries Sector

This describes the main characteristics of the fisheries sector of the country including constraints for over all development and a description of relevant sub-sector.

(c) Development and Social Objectives

This section might outline development and social objectives as expressed in national plans and official policy statements it could note the main elements of the national strategy for aquacultural development and mention significant government policies including price and interest rate subsidies supply of inputs, targets for rural income regional balance and the like.

(d) Income Distribution and Poverty

If a project is designed to benefit a particular group of the rural people, a discussion of income distribution and poverty should be appropriate. In the background section, the information should establish a framework for the eventual justification for selecting a particular region or line of action for priority attention under the project. It should cover information about income distribution on a natural basis and give a regional or social dimension to the data.

(e) Institutions Related to Fisheries Developments

Concerned with the development and financing in the fisheries sector covered by the project. These might include the Ministry of Agriculture, the live stock development authority, NABARD etc.

Project Rationale

Against the fully discussed background of development opportunities and constraints with in the relevant sectors, it should also explain why a particular development strategy has been decided for this project and establish the technical, social, and economic reasons for the selection of this particular project in preference to possible alternatives. This may be the best point at which to indicate the scale of the proposed project and to explain why a certain size has been chosen. Finally there should be a project risks and the steps that have been taken in project implementation to minimize them.

Project Area

It is to present a description of the existing status of the area, where the project will be located and to give the basis from which the project starts. These descriptive data should be presented in the relevant physical, aquacultural, social, economic, institutional and legal terms.

(A) Physical Features

This section will deal with the main geographical and topographical features of the area and relate the area to important features of the country as a whole. The principal objectives is to show that the climate and soils are suitable for the culture and live stock production proposed.

(a) *Geographic Location*

General location of the project area with in the country is identified and then the area is defined more precisely in relation to administrative boundaries.

(b) *Climate*

Should cover, rainfall, including monthly and annual total, intensity and variability, temperature, humidity, evaporation, transpiration. etc., For dry land culture the amount of water and rainfall timely is needed this is also taken into consideration.

(c) *Geology Soils and Topography*

Here take into consideration land in the project its aquacultural potential, its sustainability for culture, its need for drainage etc. Judgment will be required about the scale of the maps and land classification to be included in the report.

(d) *Water Resources*

Surface and underground water resources should be described to the extent they are relevant to project decisions. Usually this is done from the viewpoint of culture and drainage.

(B) Economic Base

This section should cover the main economic features of the project region :

(a) *Aquaculture and Live Stock Resources*

The importance of this sector is the economy of the region, the proportion of people employed in these activities, the area and an approximate estimate of the value of these products may be given.

(b) *Land use, Farming Systems and Cultures*

Include information about land type, farm size, culture type species varieties brooders, and inputs used. A short description of aquacultural practices and results achieved on experimental stations in the area may be mentioned.

(c) *Input Supply and Product Marketing*

This gives about channels for the supply of inputs and of the facilities for marketing from production. The effects of such government policies as price supports, input subsidies, taxes on products and the like may be described and evaluated.

(d) *Other Economic Activities*

These include for e.g. forestry, fishing rural handicrafts and processing industries, number of families engaged in these should be given for estimation of rural economy.

(C) Social Aspects

(a) *Land Tenure and Size of Holdings*

Land tenure should be discussed with reference to the proportion of owner cultivator, tenant cultivators and landless labourers. If possible the size of holding may be related to the kind of tenure. The descriptions should refer to any changes in land tenure caused by aquaculture reform and settlement.

(b) *Population and Migration*

Data that illustrate aspects of population as density per square kilometer, pressure of population on the cultivated area, dependency ratios and the literacy rate, migration into urban and rural and seasonal flows may be described and quantified. A discussion of employment and underemployment in the project area or near by seasonal and its relevant. Income levels discussed in the project area housing, health and nutrition of the population may be done.

(c) *Social Services*

Social services like primary and secondary schools, dispensaries and other facilities.

Disease problems their control are discussed and social services that function well and those that may need improvement are mentioned.

(d) *Infrastructure*

Depending on the project itself, infrastructure related components and requirements differ. Some projects are concerned exclusively with providing rural infrastructure in which case of course, the weight given to this section would be substantial. It may be relevant to quantify the length of road, annual tonnage; recent growth in traffic etc. infrastructure of project output and to the supply of inputs should be mentioned. The number of families served by various infrastructure facilities may be quantified.

Organization and Management

Organization and management is intended to show which entity or entities will be responsible for the various aspects of project execution and operation and how these entities will carry out their responsibilities. The discussion should demonstrate that executing agencies have adequate power, staffing, equipments and finance. It should show that adequate arrangement between and within and administrative groups responsible for various project activities. If there are deficiencies the changes and improvement required should be clearly stated.

If the administrative agency is not a government department, it may be desirable to show details about legal charts and governing board and any special provisions concerning its budget.

When there is more than one agency concerned with a project, the arrangements for coordination, joint representation on boards, joint committees and joint use of field facilities described.

It should discuss the number and caliber of the project staff whether time is enough for the operation of the project. Qualification and experience of the management staff may be noted. The needs of the project for professional and technical staff mentioned. Any necessary provision for assistance from expatriates should be noted and details about qualification of expatriates given. Some of the special requirements are:

1. credit administration
2. market structure
3. supply of inputs
4. land reform
5. research
6. extension
7. cooperatives
8. farmer organization and participation.

Production, Markets and Financial Results

The report should show that the results of project actions would be sufficiently attractive financially to encourage enough fish farmers to participate.

Production

The primary benefit of an aquacultural project is usually incremental output from project forms. This is the basis for the formulation of project. Projects may introduce new technologies but it should show the incremental yield to previous technologies. In any case the assumptions about yield or live stock production both with and without projects should be fully supported.

A table showing aggregate buildup during the development period of the project may be included in the annex.

Availability of Markets

It must be seen that satisfactory markets exists for the product of the project. It should be sufficient size to absorb the production proposed for the project, export the commodity involved, attention should be paid to such special situations as preferential treatment, long-term contracts or quality preferences.

Farm Income

Farm budgets are fundamental to any aquaculture or rural development project analysis and will also have been referred to in connection with or farm investment. It is important to present a fully developed analysis in the project report. It include farm budgets that indicate the inflow and outflow for each major farm model-anticipated in the project, out line financing needs and project the incremental net benefits the farm family may expect.

Processing Industries and Marketing Agencies

Detailed projections of balance sheet, income statement, sources and uses of funds statement and the incremental cash flow should be included.

Government Agencies or Project Authorities

In some project reports, especially if the project is to be administered by a largely self-supporting project authority, an analysis of finance from the standpoint of administering agency may be done.

Benefits and Justification

This is a crucial part of the project report in which all data discussed in previous part are brought together and an assessment made; that all things considered about whether to proceed with the project.

(*a*) *Social Benefits*

This shows the affect of income on the poorest farmers. Like :

(*a*) Income distribution

(*b*) Employment

(*c*) Access to learn

(*d*) Internal migration

(*e*) Nutrition and health

(*f*) Other indicators of the quality of life.

Some projects may have a significant effect on the quality of rural life through improvements in access to domestic water supplies, electricity, schools etc.

(*b*) *Economic Benefits*

Economic desirability of a project should be assessed. Economic cost and benefits are valued.

Outstanding Issues

All most every project will have outstanding issues that must be resolved after the preparation report is presented. These considerations may relate to project rationale policy issues affecting the project, management, staffing issues and other financing arrangements.

Sources of Institutional Assistance for Project Preparation

For specialized assistance in preparing complex projects, many governments may wish to turn to one of the bilateral or multilateral international aid agencies or to engage the services of commercial consultants.

Bilateral Assistance

Some governments may have a special interest in assisting developing countries to prepare particular projects. Information about bilateral assistance of this kind may be obtained from the embassy or equivalent office of the prospective donor country.

Examples of Bilateral Assistance

US Agency for International Development (USAID) Canada (CIDA), Denmark (DANIDO), etc., Multilateral assistance.

In addition to assistance agreed on by two governments, a government can enter into arrangements for assistance in preparing projects with a variety of international agencies.

Examples of Multilateral Assistance

Common sources are World Bank and Asian Development Bank, UNDP, FAO, etc. provide financial support to fisheries.

Contributions of multilateral agencies have been predominant till mid eighties. However, in recent years bilateral assistance has more contribution to fisheries development in indo.

World bank remains at the top among the different external sources for providing financial assistance to Indian fisheries development.

Growth of modern fisheries requires more capital advanced technology and good infrastructure, less developed countries like India would therefore depend on external assistance in order to meet the financial needs of different projects of fisheries sector.

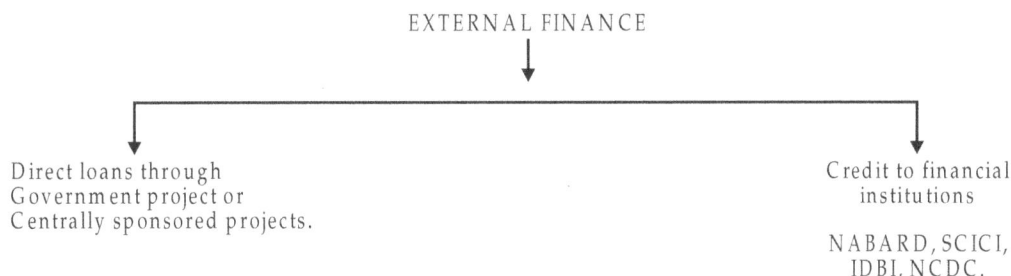

EXTERNAL FINANCE
↓

Direct loans through Government project or Centrally sponsored projects.

Credit to financial institutions

NABARD, SCICI, IDBI, NCDC.

Trends in Fisheries Financing in India

Disbursement of finance for the fisheries sector showed an increasing trend till 1995-96 after which there was a decline in the amount of credit and number of loans transacted due to following reasons :

1. Introduction of agriculture and rural financing in large scale.

2. Environmental and disease problems faced in shrimp industry.

3. Order of Supreme Court and subsequent uncertainty of final judgement on shrimp culture.

4. Slow progress in mariculture.

Biological factors	Breeding	(*a*) No breeding problem, spawning in captivity, fry available from hatchery.
		(*b*) Supply of fry relies on captured farm grown/wild gravid females, or on the catch from native waters, but availability of fry is not limited.
		(*c*) Availability of fry from natural

		waters is unlimited but the breeding problem is expected to be solved in the near future. (d) Availability of fry from natural waters is limited and the breeding problem is not expected to be solved in the near future. Sub-score or code	----------
	Feeding	(a) Nutritional requirement of different age stages are known and appropriate feeds (artificial or natural) are available at reasonable cost. (b) Nutritional requirements are partially known. (c) Nutritional requirements are not known. Sub-score or code	----------
	Crowding	(a) Adapted to crowding conditions with no major problems in diseases and parasites, and/or with wide range of oxygen tolerance. (b) Some problems with disease and parasites under crowded conditions, or with narrow range of oxygen tolerance. (c) Not suited for crowding conditions Sub-score or code	----------
	Growing period	(a) Less than one year (b) One to two years (c) More than two years Sub-score or code	----------

General score or code for resource availability

| Economic feasibility | Profitability | (a) High rate of return compared with alternatives
(b) Average rate of return compared with alternatives.
(c) Low rate of return compared with alternatives.
Sub-score or code | ---------- |
| | Socio-economic feasibility | (a) Average cost per unit of protein, or protein yield per unit of land is favorable compared with alternatives if the national policy concerns animal protein deficiency.

Foreign exchange earnings per unit of land or other scarce resources are favourable compared with other alternatives if national policy concerns foreign exchange earnings.

Employment per unit of land is | |

		favourable compared with agriculture activities if national policy concerns employment in rural areas.	
		Combination of all the above mentioned conditions or any two of them is favourable compared with alternatives.	
		(b) Partially favourable compared with alternatives.	
		(c) Unfavourable compared with alternatives.	
		Sub-score or code	----------

General score or code for institutional criteria			----------
Institutional feasibility	Permit	(a) Easy to get permit. (b) Difficult to get permit.	
	Conflict in use	(a) No conflict (b) Some conflict Strong conflict	
Market potential	Price elasticity	(a) Elastic demand with ready market (b) Inelastic demand but elasticity can be improved by market promotion and product development (c) Inelastic demand with limited market Sub-score or code	----------
	Income elasticity	(a) Elastic demand (b) Unitary elasticity (c) Inelastic demand Sub-score or code	----------
	Competition	(a) No competition (b) Competes favourably in price with close substitutes (c) Competes unfavourably with close substitutes at the present but favourably in the future (d) Cannot compete with close substitutes Sub-score or code	
	Culture, religion and tradition	(a) Species is currently cultured and preferred by majority of population without socio-cultural limitation. (b) Species is accepted by a part of the population due to socio-cultural limitations. (c) Species is not accepted due to religion or tradition Sub-score or code	----------
General score or code for resource availability			----------

Chapter 8

ORGANIC SHRIMP FARMING AND SUSTAINABILITY

Organic Farming and Organic Aquaculture

Traditional organic farming systems rely on ecologically based practices such as cultural and biological pest. Management, and virtually exclude the use of synthetic chemicals in crop production and prohibit the use of antibiotics and hormones in livestock production. Sustainability, environmental stewardship, and holistic integrated approaches to the production are hall mark of organic systems. Standards for organic cropping and terrestrial livestock husbandry practices have existed for decades.

Only if the 1990s, in response to the negative environmental and social consequences of modern methods of aquaculture, interpretation of practices and standards developed for terrestial species into practices and standards relevant to aquatic species, both animal and plant was started. Thus organic aquaculture is of recent origin.

Organic aquaculture is a method of sustainable culture practices based on long term ecologically and environmentally sound practices. It aims at producing healthy, disease free fish/shrimp with out the use of any antibiotics, chemicals, hormones etc. but at the same time ensuring that the farming activity is in harmony with nature. IFOAM and other international agencies involved in organic products certification have defined organic aquaculture more elaborately and have laid down practices and standards for organic aquaculture. The term 'organic' refers to a production system and its certification is therefore primarily certification of a production method.

Several countries are already involved in organic aquaculture production. In 2003, the estimated organic aquaculture production was 20,475 tonnes (Table 1).

While trout and salmon are the main species of organic aquaculture production in Europe, shrimp and seaweed are the principal items of production in Asia.

Organic Shrimp Aquaculture

Various socioeconomic and environmental consequences of modern shrimp farming methods in most of the shrimp producing countries lead to the search for alternative sustainable methods of shrimp farming. As organic production practices are capable of generating long-term viability of the natural resources, organic shrimp aquaculture attracted the attention of the farmers in several countries. Over 4200 ha of area have already been brought under organic shrimp production in Latin America and Asia and the production from these farms amounted to 1500 tonnes (Table 1).

Table 1 : Organic aquaculture production in 2003

Country	Species	No. of certified farms	Area (Ha)	Production (MT)
Ireland, Scotland	Salmon	6	—	5000
Austria, Germany, Switzerland & UK	Carp, Trout	40	400	500
Ecuador, Peru	White Shrimp	9	2000	1000
New Zealand	Green mussel	—	—	3600
Spain	Trout, Sturgeon	1	—	500
China	Carp, trout, tuna	5	—	4640
China	White Shrimp	1	2	4
China	Seaweed	6	1100	5000
China	Razor clam	1	2	15
Indonesia, Thailand, Vietnam	Tiger shrimp	2	2200	500
Total				20745

Standards for Organic Shrimp Production

Apart from the general principles of farm management and animal husbandry standards of organic aquaculture, specific standards have been laid down for organic shrimp aquaculture. Standards jointly laid down by IFOAM and SIPPO for organic shrimp culture are widely followed in most of the countries.

Important standards specific to organic shrimp aquaculture include:

❑ Location of the farm and the method of management should not affect adversely the surrounding eco system.

❑ While locating the farm, natural vegetation already existing in the site should not be severely damaged. Especially, mangroves should not be removed or damaged.

❑ There should be sufficient distance between organic and non-organic production areas to prevent or minimize the risk of contamination.

❑ Effluent water quality should be monitored and documented regularly. Measures must be taken to minimize out flow of nutrients, suspended solids etc.

❑ Adjacent agriculture area should not be negatively influenced.

❑ Synthetic herbicides and pesticides should not be used for removal of unwanted fish.

❑ Exotic fish species are not allowed for stocking.

❑ Stocking with wild caught postlarvae and use of wild caught brood stock should be stopped within a certain prescribed period. After that, the PLs should come from domesticated shrimp.

❑ Conventional medicine should not be used for routine and prophylactic application during production of seeds.

❑ Stocking density should be kept at minimum. Aerating or heating of the pond are not allowed on a regular basis.

❑ Antibiotics, chemo-therapeutants etc. should not be used for treating the shrimp in grow out.

❐ Fishmeal content in the feed should be reduced as much as possible. All feed ingredients should come from organic source only.

❐ Feed intake should be monitored and documented carefully. In order to avoid accumulation of organic sediments by excess feed.

❐ Farm operator should ensure free access to fishermen and other persons to open waters adjoining the farm area.

Market for Organic Shrimp

Currently, the important markets for organic shrimp are mainly Europe, but also US and Japan. The organic shrimp is sold in fresh and frozen form and distributed through specialized, health shops, organic supermarkets, and increasingly through conventional retail chains (Globe fish-2004).

In Europe, Germany is the most important market for organic shrimp. Organic shrimp products are distributed in Germany through specialized health stores, major grocery store chains, and the catering outlets. Despite price premium, demand for organic shrimp is increasing steadily in this market.

Markets for organic shrimp are increasing in other European markets also. Organic shrimp accounts to 80 per cent of the total shrimp sale of COOP, a Swiss retail chain store and a market leader for organic foods. COOP purchases the shrimp directly from a group of producers in Vietnam (supported by SIPPO) to keep prices at an acceptable level to the final consumers.

The price premium for the organic shrimp is very attractive. According to German Certifier Naturland (Globe fish 2004), in 2003, the price premium in the export market ranged from 47-62% (US $ 10.9 per kg for 41/50 headless organic against US $ 7.4 for conventional shrimp) and about 56 per cent in the retail markets in Germany. Current reports indicate that on an average the price premium is not less than 30 per cent on farm level.

Scope for Introducing Organic Shrimp Farming in India

Of the total area of over 1,60,000 ha under shrimp farming in the country traditional type of farming is carried out in about 40,000 ha. These farms are located in the states of west Bengal and Kerala.

Most of the traditional farms are big in size - ranging from a few hectares lo 30-40 ha and are tide fed. Tidal water is taken in during spring tide periods for filling the ponds. Sluice gates serve for both filling as well as for draining of water. Pumps are occasionally used. In the drainable ponds, pond preparation is limited to drying and application of lime. Apart from trapping of seeds which are carried into the ponds by tide water and holding them, additional stocking of seeds is also resorted to. Wild as well as hatchery produced seeds are procured and stocked. The stocking density is kept at a minimum - less than 1 no/square meter.

The shrimp grows using mostly the natural food available in the pond. Very rarely feeding is resorted to.

The trapping cum holding practices allow the entry of other fishes and shrimp into the ponds. They too grow along with stocked shrimp and harvested periodically. In these traditional systems, no chemicals or antibiotics are applied.

Shrimp are caught periodically by traps or through filter bags fixed in the sluice gate. Multiple stocking and periodical harvesting is practiced in West Bengal. In the pokkali fields of Kerala, paddy and shrimp are grown alternatively. These paddy fields do not apply any pesticide during paddy cultivation.

The production of shrimp from traditional farming ranges from 150 kg to 500 kg per ha per year.

The farm management practices followed in such traditional type of farms is almost meeting the standards for organic shrimp aquaculture. Several thousand hectares of contiguous farming areas in the state of West Bengal and a large number of individual farms in the state Kerala can be easily converted for organic production. The falling prices of shrimp in the international markets have brought down the profit margins considerably. Shrimp farmers will be happy to adopt organic farming if there is a scope for increasing their margins.

There is an urgent need to organize the farmers to convert their farms for organic production. Government's involvement and financial support will be essential to introduce the organic farming. Government has to take the help of an international certifier to study the farming practices followed in the two states and prepare specific management practices suiting to different farming areas and methods of farming for adoption by farmers. The procedure for the certification contract has also to be laid down. With the help of the certifier, the state fisheries departments of the two states should train their extension staff in organic management practices, documentation, monitoring of farming operation etc. These extension workers will in turn assist the farmers in providing all the information necessary for conversion to organic and in the implementation of the management practices for organic aquaculture in the farmer's ponds. They will also help to the farmers in the documentation of the details of production system.

The traditional shrimp farming practices followed in the country for over a century have proved sustainable. They are a low cost method of production using natural fertility of the ponds and thus in tune with the nature. There is no doubt that adopting organic farming will further guarantee the long-term viability of the traditional farming system of the country.

Conclusion

Returns from shrimp farming continue to be rewarding benefitting small scale farmers and communities as well as entrepreneurs engaged in seed production, farming operations or ancillary activities. Sustainable utilization of available areas and infrastructure can lead to the development of unexploited resources with the prudential of generating a large numbers of jobs and enormous social and economic benefits to the coastal regions of country.

Chapter 9

PREPARATION OF PROJECT ON ESTABLISHMENT OF SHRIMP HATCHERY

Need for the Project

Aquaculture has come to occupy a crucial place is fisheries development. The reason for this is that they have become the minimum source of seed supply for aquaculture, for rairing brooders, for induced breeding and ranching/revival of depleting wild stocks. The sudden upsurge of demand for fish and crustacean seed in the past two decades have resulted in a situation that government hatcheries are finding difficulty in catering to the requirement. This is where the importance of private hatcheries is realised.

In Kerala at present there are just seven Tiger/Freshwater Prawn Hatcheries – two in government sector (BFDA shrimp hatchery, Matsyafed Prawn Hatchery both at Kollam) and five in private sector. This proposal for a shrimp hatchery in Azhikode is desirable for uninterrupted seed supply in Kerala. As a case study, project proposal on establishment of a shrimp hatchety in Kerala is presented below.

Area

The land requirement for this hatchery is 5 ha which is taken on lease. The site may be located near the water resources. This plan has special importance. The proposed site is located near sea bar mouth the seawater could be drawn from adjoining Sea.

Profile of the Area

Climatic Conditions

The area has a warm climate (25–32°C), which is most suitable for establishment of fish hatchery since a number of seed production cycles can be obtained over an extended period.

Raw Material Availability and Utilisation

Kerala is bestowed with many good landing centers for marine fisheries. Thoppumpady is one of the largest landing centres in Kerala. It contributes 40-50% of the total fish landings. The sea catch includes a high share of pomfrets, ribbon fishes, seer fishers, perches, and several other exportable varieties such as cephalopods. The major target species of shrimps are mostly deep sea prawns, *Penaeus indicus, Metapenaecus dobsoni, Parapenaeopsis stylifera, P. monodon*, etc.

Mostly fishes are exported in the form of fillets, steaks, gutted and whole and shrimps as headless shell on and head on shell on.

A wide spectrum of raw materials like fish species, prawns and shrimps, cephalopods like cuttle fish, squid and octopuses can be processed and frozen depending on the landing trends and demand in the overseas market.

Infrastructure Facilities

The plant has accordingly been designed to incorporate three freezing systems (*i*) tunnel, (*ii*) plate and (*iii*) IQ freezing. Tunnel freezer has the capacity of 10 t, plate freezer of 5 t capacities and IQ freezing system of 2 t capacities. It possesses a flake icemaker of 5 t capacity, and a shrimp grading unit. Two cold storages of 300 t total capacity and 3 refrigerated trucks they possess. Also a well-advanced QC Lab is also there.

Water Availability

This plant depends on Municipal Corporation for its water supply. The water supplied by Kerala Water Authority need to be purified by effective chlorination system. They possess two storage tanks of capacity 100 liters each.

Their quantities for different processes of freezing

Freezing Process

Tunnel	*Plate*	*IQF*
Pomfrets (150t)	HLSO shrimp (250 t)	HLSO shrimps (250 t)
Ribbon fishes (180 t)		HOSO shimps (500 t)
Seer (90 t)		
Perches (40 t)		
Misc (40 t)		
Sub-total (500 t)	(250 t)	(570 t)
Total		(1500 t)

Staffing details and cost

Staffing	*Unit*	*Unit cost (Rs. '000)*	*Total cost (Rs. '000)*
Plant Manager	1	40	40
Procurement Asst. Manager	2	35	70
Technicians	5	30	150
Microbiologist	1	35	35
Admn./Accounts Assistants	4	30	120
Chief Supervisor	4	30	120
Supervisors	8	20	160
Workman	20	10	200
Sub-total			**895**
Add 40% (PF, bonus, gratuity, etc.)			382
Total			**1,277**
			i.e. Rs. 13 lakhs

ECONOMIC VIABILITY (INDICATIVE COST STRUCTURES/PROJECTS)

Working Capital

(Cost of unicentory for 25 days inclusive of raw material and gimsted product unicentory) (299 + 185)

Rs. 484 lakhs

	Quantity	Unit cost (Rs.)	Total cost (Rs. in lakhs)
Total Investment			
Source of Investment			
- Borrowed	60%		290.4
- Equity	40%		193.6
- Invest on borrowed capital	16%		77.44
Annual Returns			
Fishes			
- Pornfrets	100 (t)	1,00,000	100
- Ribbon fishes	100 (t)	50,000	50
- Seer fishes	200 (t)	45,000	90
- Perches	50 (t)	60,000	30
- Misc.	50 (t)	30,000	15
Sub-total	500 (t)		285
Shrimps			
- HOSO IQF	500	1,90,000	950
- HLSO IQF	250	2,10,000	525
- HLSO block	250	1,80,000	450
Total Turn Over	1500		2,210
Gross Contribution			
Turn Over			2,210
Less: Operating Cost			1,849
Gross contribution			361
Net Profit			
- Gross contribution			361
- Less: Insurance			14.60
- Less: Maintenance			
* Land and building (5%)			3.36
* Machinery and equipment (6%)			13.48
- Less: Depreciation			
* Building (5%)			2.99
* Machinery and equipment (10%)			22.46
- Less: Interest			45.76

Net Profit before tax Rs. 258.35, i.e., Rs. 258 lakhs.

Economic Analysis (*Project for 5 years*)

Discounted Day Back Period

Years	Net cash flow (Rs. in lakh)	Discounted rate (12%)	Present value of future money (Rs. in lakhs)	Cumulative value (Rs. in lakhs)
1	258	0.8929	230.3682	–
2	256	0.7972	211.258	441.6262
3	273	0.7118	194.3214	635.9476
4	281	0.6355	178.5755	814.5231
5	286	0.5674	162.2764	976.7995
Total				**976.7995**

Discounted Pay Back Period	=	By the 1st year itself.	
Net Present Value	=	Present value of future money – Initial net investment	
	=	976.79 – 477	
	=	499.79	

Internal Rate of Returns

Table given below shown the present value of cash flows discounted at two rates (one positive and other negative).

Years	Net Cash Flow	Lower Discounting 12% Rate	Present Value	Higher Discounting 49%	Present value
1	258	0.8929	230.37	0.6711	173.14
2	265	0.7972	211.26	0.4504	119.36
3	273	0.7118	194.32	0.3023	82.53
4	281	0.6355	178.58	0.2029	57.01
5	286	0.5674	162.28	0.1362	38.95
Total			**976.79**		**470.99**

$$IRR = 12 + \frac{976}{506} \times 37 = 83.36\%$$

$$\text{Discounted Benefit Cost Ratio} = \frac{976}{477} = 2.05$$

Conclusion

Since the IRR is more than the Bank rate of interest and DBCR is more than one, so the project is feasible.

Net Present Value

Year	Capital (lakhs)	Operating Cost (I)	Total Cost (I)	Cash inflow	Net cash inflow	Discounted rate 12%	Present value of cash flow
1	85.4	14.56	99.96	119	19.04	0.8929	17.00
2		60.00	60.00	100	40.00	0.7972	31.88
3		60.00	60.00	100	40.00	0.7118	28.47
4		60.00	60.00	100	40.00	0.6355	25.42
			Investment NPV				**102.77**
							− 85.50
							+ 17.37

Internal Rate of Return

Year	Net Cash Inflow	Discount 12%	Present value of cash flow	Discount 21%	Present value of cash flow
1	19.04	0.8929	17.00	0.83	15.8
2	40.00	0.7972	31.88	0.68	27.2
3	40.00	0.7118	28.47	0.56	22.4
4	40.00	0.6355	25.42	0.47	18.7
Investment NPV			**102.77**	**Investment NPV**	**84.1**
			− 85.40		**− 85.4**
			+ 17.37		**− 1.3**

$$IRR = 12 + \frac{(102.77 - 85.4)}{18.67} \times 9 = 20.4\%$$

PRODUCTION COSTS AND THEIR RELATIONSHIP

Basic Concepts

1. **Cost :** Expenditure incurred on the values and services in producing a commodity.
2. **Fixed Cost (FC)/Prime Cost :** Which does not change with the level of output.
3. **Variable/Operating/Working/Supplementary Cost (VC) :** Cost which changes with the level of production.
4. **Total Cost (TC) :** Fixed cost + Variable cost.
5. **Average Variable Cost (AVC) :** The average variable cost is worked out by dividing total variable cost by the amount output.

    AVC = VC/Y
6. **Average Fixed Cost (AFC) :** The average fixed cost is worked out by dividing total fixed cost by the amount of output.

    AFF = FC/Y
7. **Average Cost (AC) :**   = TC/Y

      = AFC + AVC
8. **Marginal Cost (MC) :** Marginal cost is the additional cost necessary to produce one more unit of output.

    Cost Curves : Typical cost curves are as follows :

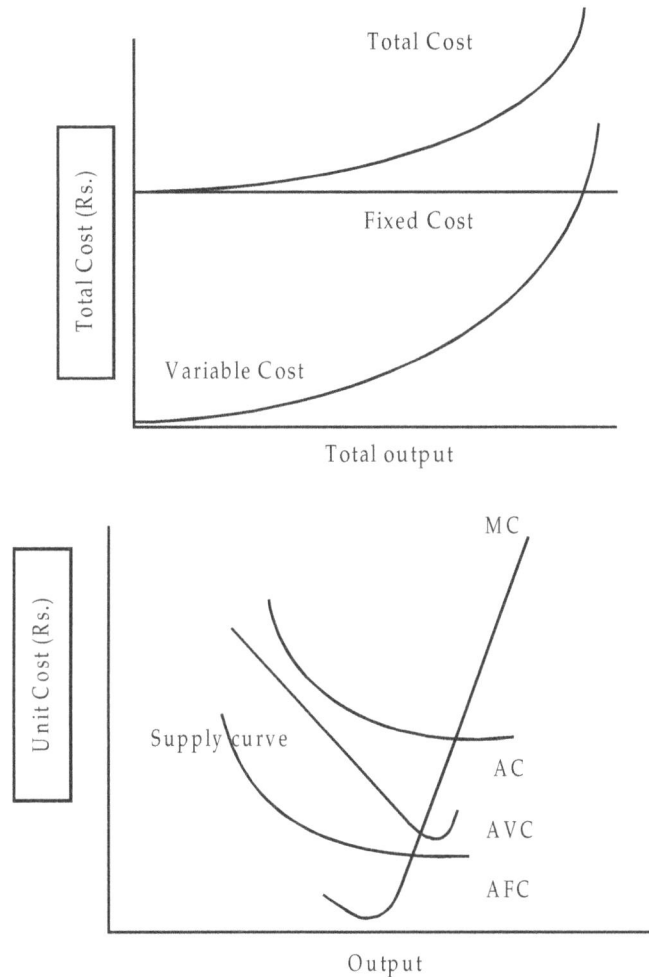

* Earliest Occurrence Time (EOT)

The earliest possible time at which the event can occur. Also denotes Earliest Start Time (EST) of an activity indicating the earliest time at which an activity commences without affecting the immediate preceding activity.

* Latest Occurrence Time (LOT)

The latest time at which the event can take place. Also referred as the Latest Start Time (LST) indicating the latest time at which an activity can begin without delaying the project completion time.

* Slack

The amount of spare time available between completion of an activity and beginning of next activity.

Steps for Network Analysis

1. Prepare the list of activities.
2. Define the presenting and succeeding relationship for all activities.
3. Estimate the activity duration.
4. Assemble the activities in the form of a flow diagram.
5. Draw the network diagram.
6. Analyse the network, *i.e.,* compute EOT and LOT; identify critical events, critical path and critical activities.

Example

Step 1 : *Prepare the list of activity and give Codes*

Each activity is given an alphabetical symbol/code. When the numbers of activities are more than 26, alpha numeric or multi-alphabet codes can be used.

SI.No.	Activity	Symbol
1.	Market Survey	A
2.	Procurement of machines for feed mill	B
3.	Installation of machine	C
4.	Selection of machine operator	D
5.	Training of machine operator with manufacturer	E
6.	Test run	F

Step 2 : *Define the preceding activities*

It gives the relationship among the activities of the project-specified by identifying preceding activities for each activity. Only the terminating activities will not have any preceding activity. All other activities must appear atleast once as a preceding activity in the table.

SI. No.	Activity	Symbol	Preceding Activity
1.	Market survey	A	–
2.	Procurement of machines	B	A
3.	Installation of machine	C	B
4.	Selection of machine operator	D	A
5.	Training of machine operator with manufacture	E	B, D
6.	Test run	F	C, E

Step 3 : *Estimation of Activity Time*

Activity time is the time that is actually expected to be expended in carrying out the activity. The expected time and its variance for each activity is computed as following:

$$Expected\ time\,(T_e) = \frac{T_O + 4T_M + T_P}{6}, \text{ where}$$

T_O -   Optimistic time   (minimum time assuming every thing goes well)

T_M -   Most likely time   (modal time required under normal circumstances)

T_P -   Pessimistic time   (maximum time assuming everything goes wrong)

Activity Table

Sl.No.	Activity	Symbol	Preceding Activity	Opti-mistic Time (T_O)	Most Likely Time (T_M)	Pessi-mistic Time (T_P)	Estimated Time (T_E)
1.	Market survey	A	–	1	2	3	2
2.	Procurement of machine	B	A	2	3	4	3
3.	Installation of machine	C	B	1	2	3	2
4.	Selection of m/c operator	D	A	2	3	10	4
5.	Training of m/c operator with manufacturer	E	D, B	3	4	11	5
6.	Test run	F	C, E	1	2	3	2

Step 4 : *Assemble the activities in the form of a flow chart*

Box — Activity and its duration

Connecting lines — According to the preceding and succeeding activity relation-ship.

Critical path for project — Comparing various path lengths (sum of activity time longest path in chart).

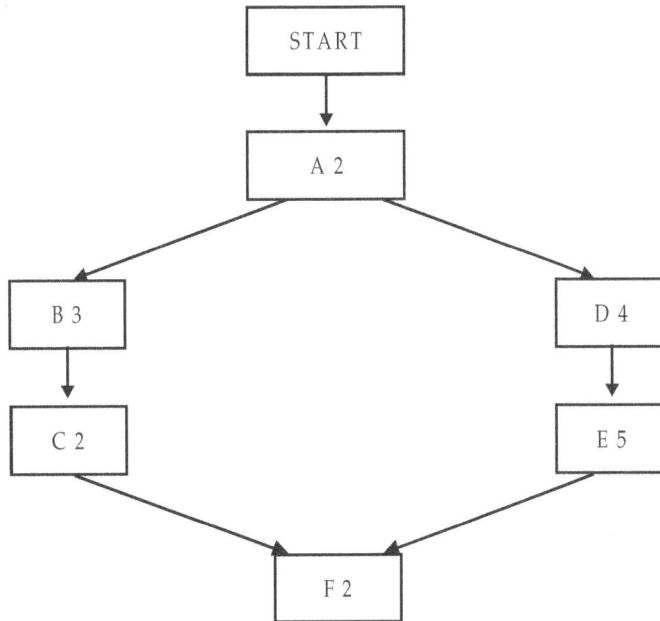

Path I	A-B-E-F	2+3+5+2	=	12
Path II	A-B-C-F	2+3+2+2	=	9
Path III	A-D-E-F	2+4+5+2	=	13

Critical path is A-D-E-F (The longest path)

Step 5 : *Draw the Network*

Rules for drawing the network.

1. Each activity is represented by one and only one arrow in the network.
2. Dotted line arrows represent dummy activities.
3. An event is represented by circle.
4. Every activity starts and ends with an event.
5. No two activities can be identified by the same head and tail event.
6. Do not use dummy activity unless required to reflect the logic.
7. Avoid looping and crossing of activity arrows.
8. Every activity, except the first and the last, must have at least one preceding and one succeeding activity.
9. Danglers, isolated activities must be avoided.
10. For coding use alphabets for all activities including the dummy activity and numbers for events.
11. Standard representation of the event.

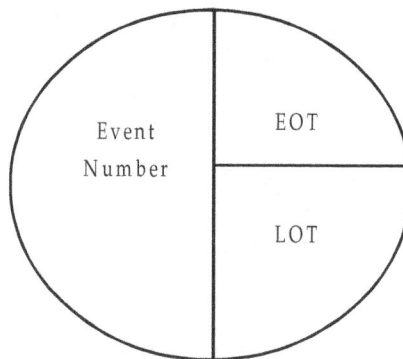

Network diagram showing the inter-relationship of activities

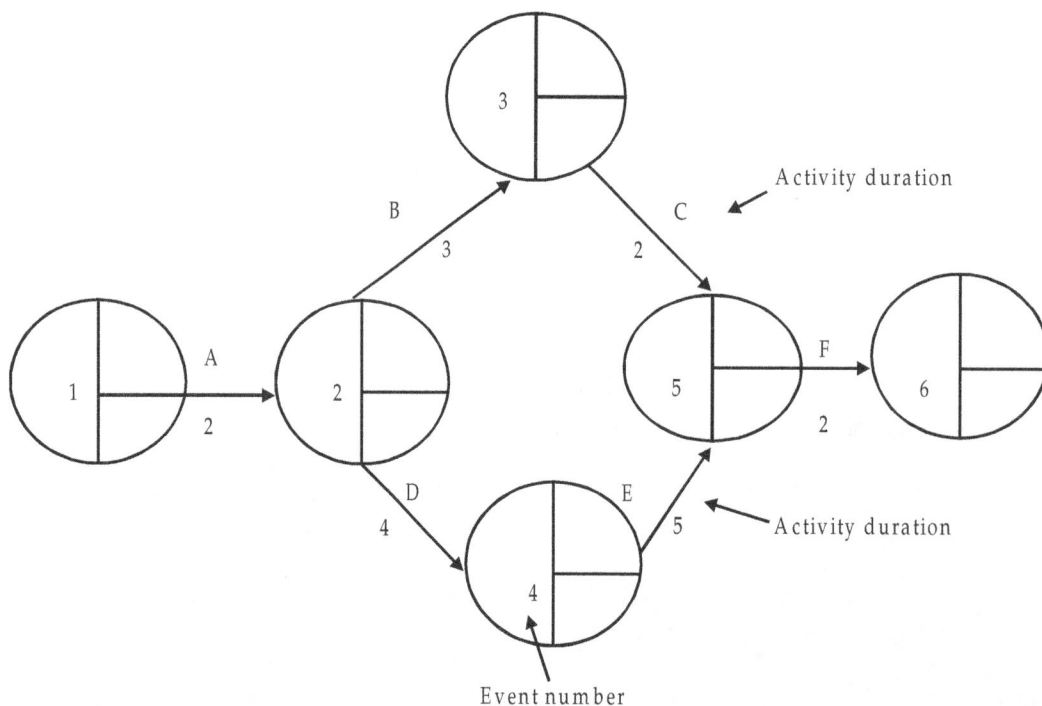

Step 6 : *Analyse the network, in compute EOT and LOT : Identify critical events and critical path and critical activities*

Computing Earliest Occurrence Time (EOT) and Latest Occurrence Time (LOT).

❑ The EOT and LOT are computed in 2 phases.

❑ The EOT is calculated first in the forward pass beginning from the start event.

❑ For the start event the EOT is always set to zero so that it can be scaled to any convenient calendar date at a later stage.

❑ The EOT at the last event is generally considered to be the project duration is the minimum time required for project completion.

❑ EOT and LOT are equal at the end event.

❑ LOT for other events is then calculated through backward pass starting from the end event.

Setup involved in computation are :

EOT	LOT
Through forward pass.	Through backward pass.
Calculation begins from start event.	Calculation starts from end event.
Proceeds from left to right.	Proceeds from right to left.
At start event EOT is zero.	At end event LOT equals to EOT.
Adding the activity time to EOT.	Subtracting the activity time from LOT.
At a merge event takes a maximum value.	At a burst event take minimum value.

Network Diagram Showing the Inter-Relationship of Activities

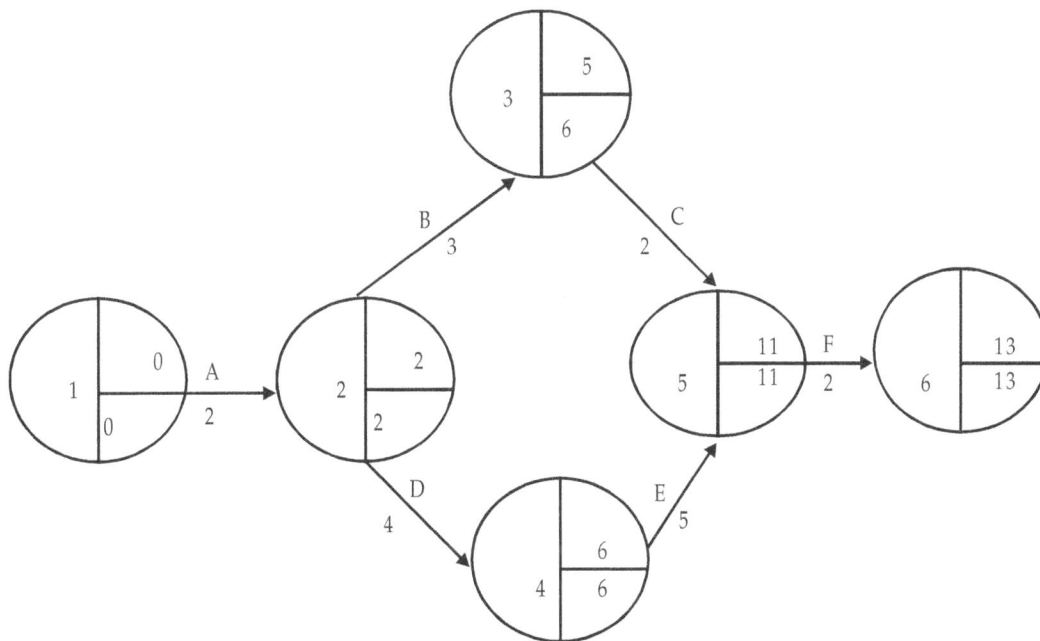

Computation of EOT and LOT for the project are as follows:

Event No.	EOT	Event No.	LOT
1.	0	6.	13
2.	0 + 2 = 2	5.	13 – 2 = 11
3.	2 + 3 = 5	4.	11 – 5 = 6
4.	Max. (2+4=6, 5+0=5) = 6	3.	Min. (6–0=6, 11–2+9) = 6
5.	Max. (5+6=11, 5+2=7) = 11	2.	Min. (6–3=3, 4–2=2) = 2
6.	11+2 = 13	1.	2–2 = 0

Identification of Critical Events

The difference between LOT and EOT for an event is called event slack. For critical events this slack is zero, i.e., the value of LOT and EOT are equal.

Event Slacks

Event No.	LOT	EOT	Event Slack	Critical/Not critical
1	0	0	0	Critical
2	2	2	0	Critical
3	6	5	1	Not critical
4	6	6	0	Critical
5	11	11	0	Critical
6	13	13	0	Critical

With above values of EOT, LOT and event slack the Critical Events are 1, 2, 4, 5 and 6.

Chapter 10

SHRIMP FARMING IN INDIA

While production from capture fisheries around the world has stagnated, aquaculture is viewed as a strong option to increase fish production, and play a vital role in providing food and nutritional security. However, aquaculture development, especially shrimp farming has been strongly opposed by environmental groups in many other countries around the globe. In India, legal actions have been taken to curtail shrimp culture and the Supreme Court's judgement lead to the setting up of an authority to regulate shrimp farming in the coastal areas or the country. Though the polarisation of opinion on the environmental impact of aquaculture in the nineties was very strong, there are signs of more tolerance to accommodate diverse views and opinions lately.

Development of Shrimp Farming – Some Issues for Consideration Commercial shrimp farming started gaining roots during the mid-eighties. It was a relatively late start in India; by this time, shrimp farming had reached peak in most of the neighbouring Asian countries, especially China and Taiwan. The boom period of commercial-scale shrimp culture in India started in 1990 and the bust came In 1995-96, with the outbreak of White Spot Disease (WSD). The fact that some coastal States in India were new to commercial-scale shrimp farming, the general ignorance of good farming practices, and the lack of suitable extension services, led to a host of problems.

It must be admitted that the days of production-oriented shrimp farming are gone. The present day production has to take note of not only the markets but a host of technical issues as well as the concerns of the environment. The subject matter of sustainable shrimp farming is broad from farm level management practices to integration of shrimp farming into coastal areas management, shrimp health management and policy, socio-economic and legal issues. In this analysis, some of the most important issues that need to be addressed by the sector on a priority basis are discussed.

Shrimp Health Management

The White Spot Disease (WSD) has played havoc and its repeated occurrence continues to be a major concern for shrimp farmers in the country. Over the last couple of years, since the spread of the WSD, many research initiatives have produced protocols, which can inhibit virus replication and also improve the resistance of shrimp to disease. The introduction of modern diagnostic tools such as Polymerase Chain Reaction (PCR) techniques to check the presence or absence of white spot virus in the shrimp post larvae prior to purchase or stocking has helped reduce the risk. Other protocols relating to pond treatment and cleanliness, and bio-security are additional developments, which considerably reduce the chances of WSD spread. In addition, new management techniques such as bio-remediation through various microorganisms and enzymes, probiotics and immuno-stimulants added to the feed are also proving to be useful and are strongly advocated.

Field laboratories with adequate facilities are essential to provide first hand diagnosis and need to be set up in the coastal. States Elaborate and comprehensive guidelines should be prepared on sustainable shrimp farming and translated in vernaculars so that the beneficiaries can be properly educated. This would help to mitigate the problems, especially those related to disease and health management.

Quality Seed and Shrimp Brood Stock Development

The availability of quality seed is likely to be a major problem in the coming years, especially with respect to the availability of brood stock. Efforts have to be made to develop programme of brood stock development at the national level. Significant advances in domestication, selective breeding, and stock improvement in recent years have been made, especially in the Western Hemisphere. The domestication and selective breeding is slow in the Eastern Hemisphere, particularly in India where no significant move has been made so far in this direction.

As the economic benefits of bio-security and genetic improvement become more compelling, R & D in selective breeding programme that rely on specific-pathogen-free (SPF) stock need to be taken up by the research institutions and also the industry to gain self sufficiency in this vital area. This paradigm shift will require dedication and cooperation from all concerned.

Feed and Feed Management

The development and use of compound feeds has been a major advancement in the successful expansion of shrimp farming. As in most animal-production enterprises, feed accounts for the largest operating cost and proper feed management is crucial for profitability of shrimp farming. Feed management techniques are as important as feed quality in both improved traditional and extensive shrimp farming. The best shrimp feed will be at best an expensive fertilizer if not managed properly.

The shrimp farming sector has received criticism in recent years for excessive use of fishmeal in formulated feeds. While the use of fishmeal posses no present threat to the sustainability of marine fisheries, it is important to develop fishmeal substitutes over time. In India, where *Penaeus monodon* is the predominant cultured shrimp species, feed protein levels have not been significantly reduced. Today, the protein levels are average around 38 per cent.

Good Management Practices

Of the many good management practices (GMPs) that are currently in vogue and adopted by the farmers, low stocking densities have proved to be successful in attaining sustainability. Based on the directives of the Aquaculture Authority, it is reported that high percentage of shrimp farms are stocking at low densities in and enjoying a high success rate for doing so. In terms of economics also, the low stocking densities are working well. There is considerable potential for India to capitalise on this position and to become an international leader in the supply of shrimp produced using environment friendly and socially responsible methods.

The GMPs for shrimp farming incorporating responsible practices are now part of the Rules of the recently established Coastal Aquaculture Authority. The GMPs extensively draw from national, regional and international experiences. These, among others, include effective and holistic farm management practices; requirements for registration of hatcheries and feed mills, community involvement, training and education, prevention etc.

The fast pace of development in the shrimp farming sector has brought to focus the use of a wide variety of drugs, chemical, antibiotics, probiotics, etc. by the fin and shellfish farmers.

Chemicals and drugs used in aquaculture include those associated with structural material, soil and water treatment, antibacterial agents, therapeutants, pesticides, teed additives, anaesthetics, immuno-stimulants and hormones. The use of most chemicals and drugs in aquaculture, if carried out properly, can be regarded as wholly beneficial with no attendant adverse environmental effects or increased risks to the health of aquaculture workers. However, the indiscriminate use of the chemicals and drugs, especially those which are banned may incur severe penalties to the shrimp farming sector, including :

❐ International trade difficulties arising from drug residues.

❐ The potential for loss of efficacy of prophylactics/antibacterial agent.

❐ Increased demand for and complexity of effluent treatment.

A potential strategy to add value to India's shrimp products is also through certification to internationally recognised standards for both quality and good practices. At present such certification process exists for only 'organic' production and products from Ecuador, Thailand, Indonesia and Vietnam are already in the market. An examination of India's practices against the natural and organic shrimp standard shows that there is strong potential for conversion to organic production despite a number of constraints. These include the need for post larvae bred from domesticated SPF broodstock, total disuse of drugs, chemicals, etc. If successful, a firm gate 'organic premium' of around 20-30 per cent can be expected.

Integrated Coastal Zone Management

The number of shrimp farms and area under cultivation has expanded considerably during the last ten years. In many areas, shrimp farms have developed in close proximity to one another (in clusters) along the creeks and estuarine watercourses. Sustained development of shrimp culture relies on good-quality source water and over-development of shrimp farms - either through management intensification or increased farm area - along a creek can impact estuarine water quality to levels unacceptable for shrimp farming. Such developments can also have impact on other users of the coastal resources, such as agriculture, forestry, etc. Therefore, this brings to focus the need for Integrated Coastal Zone Management (ICZM) programme for sustainable development of shrimp aquaculture

Aquaculture development is just one among the various developmental activities of the coastal zone and, therefore, there is need for coordination of all the developmental activities in the zone through an ICZM programme with appropriate legal and administrative support so that conflicts are the least and sustainable benefits the maximum. The role each sector has to play should be decided on the basis of the importance of the sectoral activity. As far as possible the causes for the conflicts should be removed or minimised, but in cases where conflicts still persist methods such as Zoning (permitting the specific sectoral activity where the conflicts are the least or minimal) should be resorted to.

Zoning for aquaculture may be particularly beneficial for small-scale farmers, who can be provided with proper water supply/drainage infrastructure, thus avoiding uncoordinated development of individual farms. Criteria for the identification or designation of such zones might include, for example, existing uses, land-use capability, conservation value, demographic and social characteristics and trends, hydrographic and physiographic features, etc. To ensure that the small shrimp farms, set up in clusters, do not contribute to increased nutrient load in the environment, the Coastal Aquaculture Authority is actively engaged in formulation of guidelines for setting up of common effluent treatment units.

The potential area available in the coastal region of the country for shrimp farming is estimated between 1.2 million to 1.4 million hectares. Presently, an area of about 1,57,000 ha is under farming with an average production of about 1,00,000 metric tonnes of shrimp per year. The average productivity has been estimated at 660 kg per hectare per year. Cultured shrimps contribute about 50 per cent of the total shrimp exports. The technology adopted ranges from traditional to improved traditional within the Coastal Regulation Zone (CRZ) and extensive shrimp farming outside the CRZ. About 91 per cent of the shrimp growers in the country have a holding in between 0 to 2 ha, 6 per cent between 2 to 5 ha and the remaining 3 per cent have an area of 5 ha and above.

There are around 260 shrimp hatcheries in the country with an installed production capacity of 11 billion. About 200 hatcheries are in operation producing around 7 billion shrimp larvae. There are about 33 feed mills with an annual installed production capacity of 150000 metric tonnes. Shrimp farming provides direct employment to about 0.3 million people and ancillary units provide employment to 0.6-0.7 million people.

Role of Coastal States/Union Territories

The role of the State is paramount in the development of sustainable shrimp farming. The State Departments of Fisheries have to develop a complete sense of engendering ownership of the development process and move ahead with the shrimp farmers in a participatory mode. Motivating the farmers on the use of GMPs and awareness building has been a herculean task for which the present extension paraphernalia in the States/UTs has proved to be ineffective. This has also been compounded by the absence of qualified NGOs with experience of work in fisheries sector, especially in shrimp farming. To overcome this lacuna, it is emphasised that shrimp farmers should organise into associations or 'self help group'. The formation of 'aqua clubs' in some shrimp farming areas in Andhra Pradesh and Tamil Nadu is a significant step and the State Governments should promote and further this movement to encompass all the shrimp farmers.

Opportunities also exist to strengthen the hands of the State Governments and foster public/private/civil society partnerships particularly in extending support to producers and improving farm-level management practices. Because shrimp production is currently based on extensive farming methods, the Coastal Aquaculture Authority can ensure a development path, which is more sustainable, environment-friendly and equitable.

Chapter 11

PREPARATION OF PROJECT ON CARP-CUM-PRAWN CULTURE

Introduction

Aquaculture is the system of culture of aquatic natural environment in control condition for a targeted production. The atmosphere outdoor factors remain as such. Fish production can be enhanced in par with the demographic pressures threatening food security by improving aquaculture production as capture fisheries is over exploited.

Aquaculture sector the freshwater aquaculture is of great importance due to high demand in Indian market for the fishes and prawns.

Indian Major Carps are the most sought after fishes in the land locked Indian states that include 3 species Catla, Rohu, and Mrigal. Among prawn the giant freshwater prawn *Macrobrachine rosenbergii* is on top in culture point of view. In the cultured conditions IMC and Giant Fresh Water Prawn exhibits high rate of survival, growth rate, compatibility, yield. They grow and accept feeding habits. Polyculture is viable in Indian conditions economically and technologically.

Project Title

The proposed project title is "Polyculture of Indian Major Carp with prawn".

Prospects

Cash flow analysis for a period of 10 years.

Year	Cash in flow	CASH OUTFLOW			Net Cash flow
		Capital cost	Operation cost	Total outflow	
1.					
2.					
3.					
4.					
5.					
6.					
7.					
8.					
9.					
10.					

Net Present Value

Year	Net Cash flow (Rs.)	Discounted Cash flow (12%)	Discounted Cash flow (Rs.)	Net
1.				
2.				
3.				
4.				
5.				
6.				
7.				
8.				
9.				
10.				

Pay Back Period

Year	Capital Cost	Net Cash Flow	Cumulative Cash Flow
1.			
2.			
3.			
4.			

So payable period is 3 years

Discounted Pay Back Period (all in Rs.)

Year	Capital	Net Cash flow	Discounted Factor (12%)	Discounted Net Cash Flow	Cumulative Cash Flow
1.					
2.					
3.					
4.					

So discounted pay back period is 4 years.

Average Rate of Return

$$\text{ARR} = \frac{\text{Average annual profit}}{\text{Average investment}}$$

$$\text{Average Investment} = \frac{2,20,000}{2} + 81,375$$

$$= 1,91,375$$

$$\text{Average annual profit} = \frac{1,53,625 \times 9 + 1 \times (-66,375)}{10}$$

$$= 1,31,625$$

$$\text{So ARR} = \frac{1,31,625}{1,01,375} = 68\%$$

Internal Rate of Return

Year	Net Cash flow	Discounted Cash Flow (12%)	Present Value	Discounted Cash Flow (40%)	Present Value (40%)
1.					
2.					
3.					
4.					
5.					
6.					
7.					
8.					
9.					
10.					

Internal Rate of Return = IRR

$$= r_L + \Delta r \left(\frac{(P_v - C)}{\Delta P_v} \right), \text{ where}$$

r_L = Lower rate = 12%

Δr = Higher rate

P_v = Present value

C   = Capital cost

IRR   = 39.64% that is greater than 12%

Discounted Benefit Cost Ratio

Discounted Net Benefit Cost Ratio

$$= \frac{\text{Discounted net flow}}{\text{Discounted capital outlay}}$$

Conclusion

Form the above feasibility report it is concluded that this project is benefited to satisfy the condition that is

NPV is positive.

IRR more than bank rate of interest

BCR more than one

So this project at Binka is to be taken up.

Chapter 12

INDIAN AQUACULTURE AND EVALUATION OF INVESTMENT FEASIBILITY

Several developmental programmes were launched in India, soon after independence and introduction of Green Revolution followed by White Revolution were the most significant ones initiated with the focal point of uplifting rural and sub-urban economy, besides meeting self sufficiency in food sector. The progress in these two areas were phenomenal. In similar lines promoting Blue Revolution through Aquaculture also gained momentum while orienting the attention towards meeting protein demand, both within the country and more significantly towards export front.

Although fish culture/aquaculture is an age old tradition/practice in India, particularly in the states of Kerala, Tamil Nadu, and West Bengal, a leaping success could be witnessed only when Andhra Pradesh was dotted prominently in the Indian map of culture fisheries of major carps, especially after standardising the technology for demand feeding. Today, fish production from fresh water fish culture has crossed over two million metric tones which is made available almost totally for internal consumption, thus serving as a distinct supplement for malnutrition. But the scope for promoting fresh water aquaculture with different candidate species in different culture environments such as ponds, lakes reservoirs etc. is splendid.

During the late 1980's brackish water aquaculture started budding out, but with focussed attention towards marine exports. During the entire decade of 1990's, this sector showed magnificent progress except for meeting two major challenges of:

1. Viral disease out break in shrimp farms (and)
2. A Public Interest Litigation (PIL) filed seeking judicial intervention

In spite of these two major stumbling blocks, shrimp culture sector proved its might by its major contribution, more evidently in value terms to Indian marine exports (refer table).

In addition to playing a very vital role in earning foreign exchange to the country, progress in the brackish water aquaculture sector also coupled with the development of:

1. Rural economic up-lift as a whole mainly due to :

    (a) Providing employment opportunities in rural areas – for men and women folk — skilled and un-skilled labours.

    (b) Educated youth could find self-employment avenues – all along the coastal villages of Indian maritime States.

But one should not feel complacent with this progress/achievement, as the current scenario has heavy pressure on few cultivable species (Prawns and Scampi for exports) and major carps for internal consumption, although, there exists tremendous potential for promoting diversification in aquaculture. Therefore, the urgent need of the hour is to develop aquaculture with DIVERSIFIED SPECIES and TECHNOLOGY in diversified environment, by taking the experience of other Asian countries advanced in aquaculture. At this juncture, it is very pertinent to refer that, although India is in its prestigious 2nd ranked position (2.21 million metric tons — which includes major carps, prawns and scampi) among top countries in aquaculture production, this achievement is in no way comparable to the aquaculture production of China with 28.88 million Metric tons, utilizing most of the water bodies in different ecological environments and with diversified cultivable species. It is time for India to take a lesson from this experience.

No doubt certain programmes were already initiated by MPEDA, and the notable ones among them is promoting scampi culture, which has already gained momentum and started contributing significantly to Indian rural economy. There is tremendous scope for promoting this further by strengthening the strategic approach for effective utilization of vast potential areas now lying under-utilized such as:

1. Padasekarams (Trichur/Kuttanad) in Kerala
2. Waste sugarcane lands identified in Maharastra etc.

By standardising the technology of semi-commercial seed production in hatcheries of RGCA-MPEDA, culturing of other species and as Indian sea bass and mud crab are already in pipe line.

Cage Culture

Culturing fishes in cages is one of the well known practices very well adopted in many countries all over the world (because of several advantages, as listed) but this chapter is yet to open its pages in Indian Aquaculture Scenario!!! This situation still remains mysterious as to why in India, this culture method is not adopted and rather yet to be born (whereas in several other countries it has grown to the adolescent stage), inspite of :

1. Proven Technology
2. Economic viability is established for several species

Therefore, it is desirable to have well planned/programmed outright approach in promoting this culture practice by listing out our priority agenda.

There are mainly two types of cages :

First type can be erected in off shore waters in open seas that can float and withstand rough weather conditions.

(*a*) Float in accordance to waves rising level.

(*b*) But expensive as requires additional supporting service facilities.

Second type of cages are relatively small that can be erected any where-even in inland waters–available from holding capacity of 5Mt onwards, convenient to operate and finally economical.

Advantages of Cage Culture

(*a*) Can change the culture are a from one place to another-with same capital investment and in accordance to the need/changes in environmental conditions.

(*b*) Closer monitoring of the entire culture population but need not be by random sampling as in land based system.

(*c*) Easy to handle the animals, with minimum stress.

(d) Convenience in segregation-the culture practice as commonly followed in predatory fishes like sea bass.

(e) Removing dead ones and suspected diseased animals.

(f) Almost 100% harvest of cultured produce.

(g) Easy/better maintenance.

(h) Ideally suited for culturing in reservoirs/huge water bodies having depth of more than 4 feet where draining is not possible and harvesting the produce is a very difficult task such as in Padasekarams in Trichur/Kuttanad and water bodies of Padanna in Kannur District of Kerala and possibly even in the water logged areas which were earlier utilised for brick manufacturing by digging out the soil and so on.

Indicative Fresh Water Areas in India

(A) Reservoirs 29.07 Lakh Ha (2.9 Million Ha)

(B) Tanks and Ponds 24.14 Lakh Ha (2.4 Million Ha)

(C) Floodplain Lakes and

(D) Derelict water Bodies ----------------- 07.98 Lakh Ha

(E) Figures on rivers and canals are not taken into account.

(*Source:* Handbook of Fishery Statistics (2004) Min. of Agriculture, Govt. of India).

If India could develop just 1/10th, of the area, one can imagine its mighty contribution to the Indian economy.

Vast areas of sheltered bays in Andaman are the most ideally suited water bodies for culturing marine species and promoting cage culture in those bays is very timely as this can greatly help in up-lifting the economy of the islanders particularly so under the present circumstances of :

(a) Post-tsunami scenario that warrants to look for alternative, but prospective/viable areas of investment and returns specifically so when the potential from capture fishery resource is much limited.

(b) Far better improved infrastructure facilities now available such as almost daily flights connecting to main land.

(c) Improved telecommunication and power supply. Port Blair is now identified as custom clearing port, facilitating export of live and chilled fishes, as value added items directly to importing countries is now feasible (without bringing to main land for export-which tells upon economic feasibility of any operation).

(d) Finally can aim at desirable support from govt. authorities under Island Development Programmes, as fisheries is identified as one of the thrust areas in promoting the economical growth of the islands.

(e) From the practical experience gained in launching a pilot scale cage culture project in the open seas, it is highly preferable and suggested that to begin with installing small cages of harvestable capacity of 5-10MTs, considering the technical and economic feasibility of selected Indian cultivable species.

Starting open sea cage culture can be considered only after solving/finding solution to a major issue. As such, there exists a restriction/regulation in importing exotic species. Until the time this restriction is relaxed with appropriate regulatory mechanism, it will not be economically feasible for any entrepreneur to venture in the is sector as :

(*a*) Off shore cages are expensive, and (*b*) Involves Hi-tech management

Therefore, unless the investor realizes a minimum of US$ 6-8/Kg FOB and up for live and chilled fishes of HGG (Headless-Gilled-Gutted) economic viability of the project will/may not be attractive. This indicated unit value realization is possible only for those species such as European Sea Bass/Sea Bream that are identified in affluent markets. This requires a policy decision for permitting the investors to import fingerlings/juveniles of stockable size and brooders, though not indefinitely; but until the time our hatcheries start commercial production of seeds within the country.

However, one need not wait till such time ripens. In the meanwhile it is to start even under war footing in popularising the technology of culturing Indian species such as Indian Sea Bass/Groupers etc., in small cages both in inland water bodies as well as in sheltered bays to cater to the need of both internal demand are export market.

Share of cultured fishes in export market could be enhanced considerably, in view of the steady increase in per capita consumption of fish all over the world, mainly due to very high sense public awareness created by the electronic media through their health bulletins explaining the advantage of consuming fish over red meat.

According to the predictions made by experts, this change in food consumption pattern is expected to result in wide gap between supply and demand by the year 2020 which can be sourced out only through aquaculture and certainly not from capture fisheries, due to reasons well known to all such as hike in operational cost due to price raise in fuel/labour and other overall over head expenditure, depletions of stock due to the introduction of hi-tech crafts and gears, restriction of fishing zones, etc.

Therefore aquaculture has very bright prospects with assured market demand to the extent of several million metric tons and therefore it is time for India to strengthen its approach with focused attention, taking experience of many countries in this field to our lesson and to implement appropriately.

From the lessons learnt with the introduction of brackish water aquaculture in the country, it is felt necessary to take a few operative measures for ensuring sustained growth in this sector to avoid repetition of un-desirable after thoughts such as need for assessing "Carrying Capacity" in an identified area and so on that may emerge out at a later date as major stumbling blocks in the progress of this most viable sector. Therefore, the following is suggested to consider in most pragmatic manner, for implementation.

1. An identified Central Organization should serve as a nodal agency to work at tandem with potential investors and respective state departments.

2. Although several hundred thousands hectares of area is known as per currently available records, the primary requirement is to identify the actual potential areas, taking into consideration of the local socio-economic climate, avoid multiuser conflicts (Navigation channel for the free movements of passenger boats etc.) with fore-sighted view and to specifically ear-mark such identified areas/sites for aquaculture purposes by micro-level survey when operating in open water bodies.

3. The need for scientific support to assess the "Carrying Capacity" and to recommend suitably the maximum number of cages that can be permitted to operate in an identified area, but within a specific time frame. If there is any difficulty in doing this, permit the activity in small scale and monitor the implication in the local environment (without generalizing in other areas) and to suggest remedial measures, if need be by necessarily

constituting a high-powered technical committee with the locally available scientific man-power.

4. Considering the import of proven economically viable species, though identified as "exotic" for culture in cages with the support of national research institutes for regular monitoring, coinciding with latest scientific technological developments (e.g., Producing mono-sex red tilapia seeds) for culture, which is carried out almost in 75 countries all over the globe to cash the situation of ever growing market demand; but in India, we are yet to start any programmes in this area.

5. Total legislative support with possible documentation through revenue records on the nature of sites identified, present utility, and to lease out to potential investors on long term basis (not less than 8 years), so that the projects will become viable with the support from financial institutions.

With all these pre-cautionary initiatives, one should confidently foresee and conclude that introduction and promoting cage culture is appropriate at this golden hour of implementing diversification programmes in aquaculture will certainly add another colourful feather to the cap of Indian Aquaculture Scenario/Blue Revolution.

Shrimp/prawn production from aquaculture and its role in Indian Marine exports.

Year	Production (in MTs)	Export (Value in Rs. Crores)
1979–80	51068	211.25
1989–90	57819	463.30
1994–95	1,01,751	2,510.27
2000–01	1,13,660	3,870.00
2004–05	1,64,390	3,705.00

NOTE : In ten years period during 79-80 to 89-90 production of shrimps and exports are only from capture resources except for small quantity from Chemmeen Kettu of Kerala, Bheries of West Bengal and Ghazani lands of Karnataka and Goa for which no separate figures were readily available then. There is no significant growth in value terms also in this 10 years period.

But could see distinct increase both in production and quantum jump in value once shrimps from cultured source are added to export basket. Year 2000-01 was the land mark year fetching highest unit value realization for cultured shrimps. Subsequently, though production figures shows considerable increase during 2004-05, value realization remains more or less static indicating a warning signal to; that over dependence on single cultivable species is no more desirable under the varied developments that are taking place, such as increased production from China and other South East Asian Countries, like Vietnam which has emerged out as one of leading producers of shrimps from culture, currency fluctuations, volatile market conditions due restrictions imposed by the importing countries such as stringent quality standards, trade barriers etc. Therefore, providing emphasis on popularizing diversification in aquaculture is very timely and introducing cage culture is much more appropriate. According to latest data the fish production of the country is 6.7 mmt in total.

EVALUATION OF INVESTMENT FEASIBILITY OF AQUACULTURE PROJECTS

Basic Concepts

To as certain whether an aquaculture investment project is feasible or not, a co-operative evaluation should first be conducted by both the biologist and the economist. Only those species and projects that are suited to the local environment and are biologically feasible for development should be considered. Thereafter, a socioeconomic study can be undertaken functions. Many considerations affecting the feasibility of aquaculture investment are briefly summarized below.

1. The first requirement for any aquaculture investment project in both the public and private sectors is the availability of suitable land and water resources.

2. The selection of species for development should be adapted to the local environmental conditions and the stocking materials and suitable feed should be readily available at reasonable cost. The species should also be fast growing and culture technology should be locally available.

3. There should be no legal constraints on development for private investors.

4. The products of the investment project should have a high market demand with a reasonable price.

5. The investment project should be financially lucrative compared to other investment opportunities for private investors and should also be socioeconomically feasible with alternative means of achieving the national objectives for public investment. Private investors usually use profitability as a measure of financial feasibility when assessing commercial aquaculture projects, and public officials usually consider socioeconomic benefit-cost and/or the social internal rate of return as measures of economic feasibility along with some qualitative judgements.

To evaluate the feasibility following six criteria may be considered.

- Resources availability.
- Environmental suitability.
- Biological feasibility.
- Market potential.
- Economic feasibility, and
- Institutional feasibility.

6. (*a*) Each variable can be assessed as favourable, partially favourable, unfavourable, etc.

(*b*) Each ranking can then be scored (or coded) numerically—weighted or unweighted. Next, a general score or code can be assigned to each criterion after evaluation of all the subscores and codes, and the bio-economic feasibility can be determined by weighting the general score or code criterion.

7. Summary sheet for feasibility evaluation is given below :

Summary Sheet for Feasibility Evaluation

Criteria	Variables	Rank of Suitability		Score of Code
Resources	Suitable land area	a.	Available for expansion limited for expansion	
		b.	Not available for expansion Sub-score or code	----------
	Value of suitable land	a.	Low	
		b.	Average	
		c.	High	
			Sub-score or code	----------
	Water supply of suitable quality	a.	Adequate year round	
		b.	Seasonal shortage	

Contd...

		c. Not available Sub-score or code	-----------
	General score or code for resource availability		-----------
Environ-mental suitability	Water temperature	a. Well suited	
		b. Suited after temperature is manipulated during certain periods of the year	
		c. Not suited	
		Sub-score or code	----------
	pH value	a. Well suited	
		b. Suited after pH value is manipulated	
		c. Not suited	
		Sub-score or code	----------
	General score or code for resource availability		---------

Project appraisal techniques are two *i.e.* undiscounted and discounted. The basic difference between these two lies in the consideration of time value of money in the project investment. Undiscounted measures do not take into account the time value of money, while discounted measures do.

Many economic decisions including fish production involve benefits and costs that are expected to occur at future time period. The construction of ponds race ways, and fish tank, for example, requires immediate cash outlay, which with the production and sale of fish, will result in future cash inflows or returns. In order to determine whether the future cash inflows justify present initial investment, we must compare money spent today with the money received in the future. It is true that cost changes with time.

The time value of money influences many production decisions. In order to invest a rupee in fish production it is guaranteed to get more money return in the future. The preference for the rupee now instead of a rupees in the future arises from three basic reasons: Uncertainty, Alternative uses and Inflation.

(*a*) *Uncertainty,* influences preferences because one is never sure what will take place tomorrow.

(*b*) *Alternative uses,* it will determine whether one invests in one project or another.

(c) *Inflation,* affects the purchasing power of the rupee.

Accordingly the undiscounted measures of project worth includes :

- Ranking by inspection
- Pay back period
- Proceeds per unit out lay
- Average annual proceeds per unit out lay method

Discounted Measures of Project Worth

The technique of discounting permits to determine whether to accept for implementation, projects that have variously shaped time streams *i.e.,* patterns of when costs and benefits fall during the life of the project that differ from one another – and that are of different durations. The most common means of doing this is to subtract year-by-year the costs from the benefits to arrive at the incremental net benefits stream–the so-called cash flow-and then to discount that. This approach will give one of three discounted cash flow measures of project worth–the net present worth, the internal rate of return or the net benefit investment ratio. Another discounted measure of project worth is to find out the present worth of the cost and benefit stream separately and then to divide the present worth of the benefit stream by the present worth of the cost stream to obtain the benefit-cost ratio.

Because the benefit and cost streams are discounted, the benefit-cost ratio is a discounted measure of project worth. But benefit and cost streams are discounted separately rather than subtracted from one another year-by-year, the benefit-cost-ratio is not a discounted cash flow.

Discounted Payback Period

It estimates the length of the time required for an investment to itself out; that is the number of years required for a firm to cover its original investment from the net cash inflows.

Although the period is easy to calculate, it can lend to erroneous decisions. As can be seen from our example, it ignores income beyond the payback period, and biased towards projects with shorter maturity periods. The payback period is sometimes used by investors who are short of cash need to reinvest all cash flows that occur in early stages of the projects. Investors who are risk averse often use this technique in evaluating projects. Such investors need to receive cash at the early stages of projects since the future is uncertain. This, the payback period method is somewhat better reflection of liquidity than profitability.

Table : Net cash inflow for project A & B

	Project A					Project B			
Year	Invest-ment	Net cash inflow	Discount factor (12%)	Present value of net cash inflow	Year	Invest-ment	Net cash inflow	Discount factor (12%)	Present value of net cash inflow
0									
1									
2									
3									
4									
5									
6									
7									

Net Present Value (NPV)

A discounted cash flow technique (DCF) and present value discounted at firm's required rate of return on the stream of net cash flows from the project minus the project's net investment. The NPV method uses the discounting formula of a non-uniform or uniform series of payments to value the projected cash flow for each investment alternative at one point in time. To obtain the NPV, following formula is used :

$$NPV = -INV + \frac{P_1}{(1+i)^1} + \frac{P_2}{(1+i)^2} + \frac{P_3}{(1+i)^3} + \ldots\ldots + \frac{P_n}{(1+i)^n}$$

where, $P_1 \ldots \ldots P_n$ are net cash flows.

i = the interest rate or marginal cost of capital

n = the project expected life.

INV = the initial investment.

The model indicates that the net cash flows of the project are discounted and then added to yield the NPV. The initial investment is negative since it represents a cash flow.

or

$$NPV = \sum_{t=1}^{n} \frac{P_t}{(1+i)^t} - INV$$

An investment project would be accepted if the NPV > 0 and rejected if NPV < 0. This is because the money being invested is greater than the present value of the net cash flow. If NPV = 0, the decision maker would be indifferent. The NPV method assures that funds may be reinvested at the firm's interest rate. In case of series of cash out flows and cash inflows they can be written as

$$NPV = \sum_{t=1}^{n} \frac{B_t - C_t}{(t+i)^t}$$

where, $B_t \rightarrow$ Benefit in each year

$C_t \rightarrow$ Cost in each year

$i \rightarrow$ Discount rate

Benefit Cost Ratio (BCR)

It is also called as profitability Index (*PI*). The ratio of present value of future net cash flows over the life of the project to the net-investment.

$$PI = \frac{\sum_{t=1}^{n} \frac{P_t}{(1+i)^t}}{INV} \quad or \quad PI = \frac{\sum_{t=1}^{n} \frac{B_t}{(1+t)^t}}{\sum_{t=1}^{n} \frac{C_t}{(1+i)^t}} \quad \begin{pmatrix} \text{in case of series of cash outflows} \\ \text{and cash inflows in years} \end{pmatrix}$$

The method usually produces the same result as the *NPV* and *IRR* in project evaluation, but it is very important in separating projects of varying sizes. If a project has a *PI* value greater than or equal to 1, (*PI* = 1) it should be accepted and should be rejected if the *PI* value is less than 1 (*PI*<1).

Example : A fish culturist has invested and got Net benefit are the end 1st, 2nd, 3rd and 4th year of fish culture in the following way:

Year	Investment (Rs.)	Net benefit	Discount factor (12%)	Present value investment	Present value of Net benefit
0					
1					
2					
3					
4					
Total					

NPV = Present Value of Net Benefit – Present Value of Investment

$$BCR \ or \ PI = \frac{\text{Present value of Net benefit}}{\text{Present value of investment}} = 1.12 \ (\text{more than } 1)$$

Internal Rate of Return (*IRR*)

It is the interest rate that will equate the sum of net cash flows to the initial investment. The interest rate that satisfies the equation is called internal Rate of Return (IRR).

There is no way of finding the IRR. One is forced to use a systematic procedure of trial and error to find out the discount rate that will equate the net cash flows to the initial investment. When the NPV = 0, Then

$$\sum_{t=1}^{n} \frac{\frac{Bt-Ct}{(1+i)^t} Pt}{INV \ or \ \Sigma} = 0 \quad \text{(in case of series of cash flows)}$$

$$t = 1 \quad (1 + i)^t$$

i = Internal Rate of Return (*IRR*).

Acceptability of project depends upon comparing the IRR with the investor's required rate of return (RRR) sometimes called minimum acceptable rate of return (MARR). If IRR is greater than RRR (MARR), accept the project, if IRR is less than that, reject the project, if IRR=RRR, be indifferent.

If NPV is greater than (or less than) zero (0), and only if the IRR is greater than (or less than) RRR, the NPV and the IRR method result in identical decisions to either accept or reject an independent project.

The IRR method implicity assumes that returns from an investment are reinvested to earn the same rate as the IRR of interest.

Example : Initial investment capital for composite fish farming is Rs.

Year	Cash flow (Rs.)	Discount factor (12%)	Present value (12%)	Discount factor (20% Rb)	Present value (20%)
0					
1					
2					
3					
4					

NPV (for 12 per cent discount rate Ra) = Discount factor (Rb 20%) – Capital Investment

NPV (for 20 per cent discount rate Rb) =

$$IRR = Ra + \frac{(Pv - V)\ddot{A}r}{DPv}$$

Where, Ra = Minimum rate of interest (i.e. 12 per cent as bank rate of interest)

Pv = Present value of cash flow (at Ra)

C = Capital

DPv = Difference between the present values

= NPV (Ra) – NPV (Rb)

Är = Rb – Ra

The rule for the selection of the project is :

DPBP = minimum

NPV   = greater than 0

BCR   = more than 1

IRR   = more than the bank of interest

Risk and Uncertainties in Projects

One of the real advantages of careful economic and financial analysis in fisheries project is that it may be used to test what happens to the earning capacity of the project if events differ from guesses made about then in planning for *e.g.* a disease outbreak, fall in prices, natural calamities like floods, etc. These unforeseen incidents can be grouped into two *i.e.* risks and uncertainties.

Risk

Few management decisions are made under conditions where the outcomes associated with each possible course of action are known with certainty. Most major managerial decisions are made under conditions of uncertainty. The frequency of uncertainty in managerial decisions and the risk involved dictate risk analysis be given due consideration in farm and project management decisions. Risk refers to the possibility that some unfavourable event will occur. It is the possibility that some unfavourable event will occur. It is the possibility of loss, injury, or exposure to harm. In aquaculture, risk comes from stock losses. Anything, which disrupts the rearing of fish, in likely to jeopardize production and marketing of the final product.

The levels of risk vary among species and at different stages of production. The relative lack of knowledge of fish biology in comparison to some land animals makes fish production more risky than the production of food animals. As Secretan (1988) indicates, on a scale of 1 to 100, 20 per cent of the biology of aquatic species. There are numerous risks involved in the breeding, hatching and growing of aquatic organisms under intensive management systems.

What sort of risks plagues the aquacultural industry? Risks may be classified into main groups: (1) Socio-economic or business risk and (2) physical or pure risks.

TYPES OF RISK

Socio-Economic/Business

Social aspects of socio-economic risks include changes in tastes, attitudes, or social behaviour towards production an consumption of a certain species. The expansion of aquaculture depends on individuals changing their attitudes towards species cultured under intensive closed systems. This may be done through government programs, advertising, and public relations. For example, changes in consumer purchases of ornamental have been achieved through advertising and public relations. The growing popularity of ornamental may be satisfied, however, if "of flavour" in colour problems continue to plague the industry.

Economic Risks

Economic risks such as changes in price of inputs and output, recession, depression and other economic conditions which affect national income are primary concerns of commercial fish producers. As demand lags behind supply producers are concerned that prices will fall. This is presently the case in the every where ornamental fish industry. Producers are being informed they should secure markets before expanding production. A clue to the level of economic risks associated with fish production. Further marketing facing a more inelastic demand than producers will tend to be less concerned about demand lags. This is one reason that producers are beginning to favour more producers associations or co-operative type marketing organizations.

Marketing Risks

Risks may also result from uncertainty in demand, supply and prices. When to move to market is the age-old nemesis of farmers. Fish farmers are no different. Significant seasonal price level differences exist in many aquaculture product markets. New technologies are being evaluated in an attempt to avoid some of the marketing risks. Hardy indigenous fishes used in various markets to reduce the risks associated with marketing time.

Assume that forecasters are overly optimistic in their estimates of prices and consumer demand. This optimism is likely to encourage farmers to intensify production (higher stocking rates) in the short-run and expand production (more ponds) in the long run. Intensification increases the potential for diseases, problems such as "off-flavor in colour", and other environmental concerns. The fish arrive at the market only to remain unsold because of weak consumer demand resulting from a dislike for the quality of the fish on the shelf, or insufficient income to purchase fish and other market material. The most important thing is unawardness.

Longer-term expansion of production means greater amount of capital and land committed to the aquacultural practice. Because ponds are much easier to build than to remove, these commitments tend to become irreversible, even if prices decline. Once again market conditions dictate many difficulties for the producer.

Production Risks

Many of the marketing risks are also related to production problems. Marketing problems may be logistical in nature, which may impede production schedules. The timely supply of fingerlings may affect the quantity of food fish produced at a given time. This may result in grave financial problems for producers. Production risks may also be due to lack of trained manpower to manage the operation. This results in serious constraint or even failure in any aquacultural enterprise.

Other Risks

Other socio-economic risks encountered are financial and political. Financial risks relate to changes in supply of funds for production and marketing. Credit restriction and availability often affect the aquacultural industry. Lack of education and understanding of aquacultural production processes among lenders is common in areas where the industry is developing.

Political risks affect not only an enterprise, but the whole sector. Changes in government and governmental policies have been known to cause changes in supply and demand of inputs fish. Governmental regulations and may affect all stages and aspects of the industry. Regulations on feed, import of inputs, the introduction of species, and changes in labour laws may greatly influence the industry.

Physical or Pure Risks

Physical risks results from conditions of nature, such as rain windstorms, clouds, flooding, and drought. Other types of pure risks are plant breakdowns, and failure of safety and other devices. These risks associated with physical or pure risks can be managed to minimize their effects on producers.

Uncertainty

It is a situation in which the probability of an outcome is not known. Insurance cannot provide any cover against uncertainty. Uncertainty is a state of being doubtful about future events, which cannot be foreseen exactly.

Types of Uncertainty

❑ **Price uncertainty :** It is associated with the price of products and input factors, such as price of fish in a market.

❑ **Yield uncertainty :** The fluctuations in yield are associated with weather conditions and incidence of diseases and pests and the impact of new practices.

❑ **Technological uncertainty :** Technological changes influence production function and create conditions of variability, which, in turn, lead to uncertainty.

❑ **Institutional uncertainty :** Conditions of tenure, functioning of credit agencies, action and outlook of farmers are examples of institutional uncertainty.

Normally risks and uncertainties are removed by the following methods:

(*a*)   Diversification

(*b*)   Crop insurance

(*c*)   Continuous or Sequential Marketing

(*d*)   Future Market or Production Contracts

(*e*)   Government Programs

(*f*)   Third-Party Equity Capital

(*g*)   Use of Safety Device

The risk and uncertainties in the fisheries projects can be accounted by the following methods.

* Sensitivity Analysis

Sensitivity analysis is a simple technique to assess the effects of adverse changes on a project. It involves changing the value of one or more selected variable and calculating the resulting changes in NPV or IRR. The extent of change in the selected variable to test can be derived from post evaluation and other studies of similar projects. Changes in variables can be assessed one at the time to identify the key variables. Possible combinations can also be assessed. Sensitivity analysis answers questions like what happens to NPV if the sales of the output are 10 tons rather than the expected 15 tons? What will happen to NPV if the economic life of the project is only 6 years rather than the expected 8 years? How sensitive is the projects financial and economic rate of return or net benefit investment ratio to increased construction costs etc.

Where the project is shown to be sensitive to the value of a variable that is uncertain, mitigating actions should be considered. This can include project level actions, such as long-term supply contracts or pilot phases; sector level actins, such as price changes or technical assistance programmes; or national level actions, such as changes in tax and incentive policies, where there is exceptional uncertainty, the project may have to be redesigned or implemented first on pilot basis.

Sensitivity and risk analysis can be used to assess the effects of changes in project variables that are quantified. Many projects involve institutional and social risks that cannot be readily quantified. A statement of such risks and any mitigating actions should be included alongside the conclusions from sensitivity and risk analysis.

Merits of Sensitivity Analysis

1. It forces management to identify the underlying variables and their relationships.

2. It shows how robust or vulnerable a project is to change in underlying variables.

3. It indicates the need for further work in terms of gathering information if NPV or IRR is highly sensitive to changes in some variable.

Generally projects are sensitive to change in 4 principal areas. These and the technique of sensitivity analysis are considered below :

* Uncertainties of Price Receivables

Probably every aquacultural project should be examined to see what happens if the assumptions about the sale price of the project's product prove wrong. For this the analyst can make alternative assumptions about future price and see how these affect the net present worth the financial and economic rates of return, or the net benefit-investment ratio (often abbreviated as N/K ratio).

* Delay in Implementation

Delay in implementation affects most aquacultural projects. Farmers may fail to adopt new practices as rapidly an anticipated or they may find it harder to master new techniques than was thought. Other technical difficulties may be underestimated. There may be delays in ordering and receiving new equipment. Unavoidable administrative problems and requirements may delay the project. Testing to determine the effects of delay on the present worth, the financial and economic rates or return and the net benefit investment ratio of a proposed agricultural investment is an important part of the sensitivity analysis.

* Cost Overrun

Almost every aquacultural project should be tested for sensitivity to cost overrun. Projects tend to be very sensitive to cost overrun especially for construction because so often the costs are incurred early in the project when they weight heavily in the discounting process and are for facilities that must be complete before any benefit can be realized. A project that has a quite attractive return if the estimated cost is in fact realized may be only marginally acceptable or unacceptable if costs early in the implementation phase rise significantly.

Cost estimates often are not very firm is one more reason why projects should be tested for cost overrun.

A test that shows a project to be very sensitive to cost overrun signal to those who must make investment decisions that it is important to have firms cost estimates before proceeding with the final decision, even if obtaining firm estimates may mean a delay in the start of project implementation. If a project manager and those to whom he reports that it is important to contain costs if the project is to make its expected contribution to increasing national income.

* Yield

The analyst may wish to test a proposed project for its sensitivity to errors in estimated yield. There is a tendency in projects to the optimistic about potential yields, especially when a new cropping pattern is being proposed and the agroclimatic information is based mainly on experimental trials. A test to determine how sensitive the projects net present worth, financial and economic rates or return, or net benefit-investment ratio are to lower yields not only may provide information useful in deciding whether to implement the project, but also may emphasize the need to ensure sufficient extensions services if the project is to be as high-yielding as could reasonably be experienced.

Sensitivity analysis is a straightforward (but often quite sufficient) means of analyzing the effects of risk and uncertainty in project analysis. A much more elaborate technique of risk analysis using probability [Pouliquen (1970) approach] is generally called "probability analysis". In contrast, the techniques we have been discussing (including sensitivity analysis) are usually called "most probable outcome analysis".

Problem

For the following fisheries project perform the sensitivity analysis for the three different cases of :

(*i*) Increasing cost of capital.

(*ii*) Increased cost of project due to risks involved at 10 and 20 per cent cost like.

(*iii*) Uncertainties due to the differences in the receivables at 10, 20 and 30 per cent reduction for the yield.

Sensitivity Analysis

Case I : Increasing Cost of Capitals

Year	Cost	Benefit	D.F. 12%	D.C. 12%	D.B. 12%	D.F. 20%	D.C. 20%	D.B. 20%	D.F. 25%	D.C. 25%	D.B. 25%
1											
2											
3											
4											
5											
6											

Inference

The computation of the NPV and BCR at different cost of capital indicates that the project is feasible and profitable even at 25 per cent discount rate. At 25 percentages discount rate also there exists a positive NPV and BCR of more than one. The exercise indicates the high yielding capacity of the project even at higher discount rates.

Case II : Escalation of the cost of the project due to the different risks involved

Year	Cost	Benefit	D.B. 12%	D.C. 12%	D.B. 12%	Cost increase by 10%	D.C. 12%	D.B. 12%	Cost increase by 20%	D.C. 12%	D.B. 12%
1											
2											
3											
4											
5											
6											

Inference

On increasing the cost of the project taking into consideration the different risks involved the computed NPV and the BCR values indicate that the project is feasible and economical upto a discount level rate of less than 20 per cent cost increase. At 20 per cent increase in the total cost of the project the NPV appears to be negative and the BCR is lesser than one which are negative indicators of project appraisal.

Case III : Uncertainties resulting due to differences in the price receivables.

Year	Cost	Benefit	D.F. 12%	D.C. 12%	D.B. 12%	Reduction in benefit of 10%	Discounted benefit	Reduction in benefit of 20%	Discounted benefit	Reduction in benefit of 30%	Discounted benefit
1											
2											
3											
4											
5											
6											

Inference

The uncertainties in the project benefit stream can be sensitized by the ex-ante approach of reducing the anticipated project benefit stream at 10, 20, 30 percentage. The computed NPV and BCR ratios indicate that the project can withstand uncertainties to the tune of even 30 per cent in the yield due to the different uncertainties. The NPV and BCR at 30 percentage reduction in the yield in the project benefit stream was found to be Rs 9429 and 1.21 respectively.

Limitations of Sensitivity Analysis

1. In may fail to provide leads if sensitivity analysis merely presents complicated set of switching values it may not shed light on the characteristics of the project.

2. The study of the impact of variation in one factor at a time, holds, other factors constant, may not be very meaningful when underlying factors are likely to be interrelated.

Case AQ

(A) Transportation cost paid by Mr. Kasim, fisherman

   $$= 100 \times 0.50 = \text{Rs. } 50$$

(B) Transportation cost paid by Mr. Rama auctioneer-cum-retailer

   $$= 100 \times 0.60 = \text{Rs. } 60$$

(C) Costs for icing paid be Mr. Rama

   $$= 100 \times 0.40 = \text{Rs. } 40$$

   $\therefore$ Total Marketing Costs (MC) = A + B + C

   $$= 50 + 60 + 40$$

   $$= \text{Rs.150}$$

Profit earned by Mr. Rama = 100 {45 − (40 + 0.60 + 0.40)}

   $$= 100 (45 - 41)$$

$$= \text{Rs. } 400$$

∴ Total Marketing Margin (MM) = Rs. 400

∴ Price spread for 10 kg Sciaenids

$$= MC + MM$$
$$= 150 + 400$$
$$= \text{Rs. } 550$$

∴ Total price received by Mr. Kasim, fisherman

$$= (100 \times 40) - 50$$
$$= 4000 - 50$$
$$= \text{Rs. } 3,950$$

∴ Total price paid by the consumers

$$= 100 \times 45$$
$$= \text{Rs. } 4,500$$

∴ Producer's share in consumer's rupee

$$= \frac{3950}{4500} \times 100 = 87.78 \text{ per cent.}$$

Chapter 13

SENSITIVITY ANALYSIS

Basic Concepts

Sensitivity analysis is a simple technique to assess the effects of adverse changes on a project. It involves changing the value of one or more selected variables and calculating the resulting change in the NPV (Net Present Value) or IRR. The extent of change in the selected variable to test can be derived from post evaluation and other studies of similar projects. It can be applied for numerically large or uncertainties.

Change in variables can be assessed one at a time to identify the key variables. Possible combinations may also be assessed.

This results to what actions to take or which variables to monitor during implementation and operation.

Merits

- ❑ It forces management to identify the underlying variables and their relationships.
- ❑ It shows how robust or vulnerable a project is to change in underlying variables.
- ❑ Gathering information in NPV or IRR is highly sensitive to changes in some variables.

Demerits/Limitations

1. It may fail to provide leads – if sensitivity analysis merely presents complicated set of switching values it may not shed light on the characteristics of the project.
2. The study of impact of variation in one factor at a time, holds other factors constant, may to be very meaningful when underlying factors are likely to be inter-related.

Methodology

Sensitivity analysis can be done to ascertain the project feasibility at three different stages.

(i) Increasing cost of capital or interest rate increases

The increasing cost of capital or the interest rate increases can be accounted in the sensitivity analysis be computing the NPV and BCR at different discount rates and thereafter checking the profitability of the changes.

(ii) Escalation of cost of the project due to different risks involved

The cost of the projects gets escalated due to the various risk factors in the business prophylactic measures, control and prevent the disease outcome, use of fertilizers, more number of irrigations, more number of man days increase due to inefficiency of labour, etc. These

increase in the ex-ante approach of increasing the project cost by 10 per cent and 20 per cent and later working the NPV and BCR with the benefit stream keeping unchanged.

(iii) Uncertainties resulting due to differences in the price receivables

The uncertainties in the project benefit stream arise due to the uncertain nature of the prices that are expected in the market after the harvests. The uncertainties are basically due the factors determining prices itself are subjected to changes. Like the yield uncertainty, technological uncertainty and institutional uncertainty. In countering the uncertainties, the anticipated benefit stream in the project can be reduced by 10,20,30 percentages and the NPV and BCR are computed accordingly, keeping the project cost unchanged.

Example

For the following fisheries project perform the sensitivity analysis for the three different cases of :

-   (i) Increasing cost of capital.
-   (ii) Increased cost of project due to risks involved at 10 and 20 per cent cost like.
-   (iii) Uncertainties due to the differences in the price receivables at 10, 20, and 30 per cent reduction for the yield.

Conclusion

1. The computation of the NPV and BCR at different cost of capital indicates that the project is feasible and profitable even at 25 per cent discount rate. At 25 percentage discount rate also there exists a positive NPV and BCR of more than one. The exercise indicates the high yielding capacity of the project even at higher discount rates.

2. On increasing the cost of the project taking into consideration the different risks involved the computed NPV and the BCR values indicate that the project is feasible and economical to a discount level rate of less than 20 percentage cost increase. At 20 percentage increase in the total cost of the project the NPV appears to be negative and the BCR is lesser than one that are negative indicators of project appraisal.

3. The uncertainties in the project benefit stream can be sensitised by the exante approach of reducing the anticipated project benefit stream at 10, 20, 30 percentages. The computed NPV and BCR ratios indicate that the project can withstand uncertainties to the tune of even 30 per cent reduction in the yield due to the different uncertainties. The NPV and BCR at 30 percentage reduction in the yield in the project benefit stream was found to be Rs. 9,429 and 1.21 respectively.

FINANCIALLY AND ECONOMICALLY

Basic Concepts

Financial prices

Actual prices at which inputs are bought sold and are used in financial analysis. Also called nominal current or market price.

Constant prices

Constant price refers to a value from which the overall effect of general prices inflation has been removed.

Shadow Prices

Shadow prices are prices indicating the intrinsic or true value of a factor or product in the sense of equilibrium prices. These prices may be different for different time periods as well as geographically separated areas and various occupants (labour). They may deviate from market prices [Tinbergen 1954].

Official Exchange Rate [OER]

This is the agreed rate of exchange between two countries. It is used in financial analysis.

Shadow Exchange Rate [SER]

The SER is the weighted average of the demand price of foreign exchange paid for by the importers and the supply price of foreign exchange received by the exporters. It is the economic price of foreign exchange.

The relation between OER, foreign exchange premium [the percentage difference between SER and OER], SER and SCF are given below :

$$\text{OER} \times [1 + F \times \text{premium}] = \text{SER, and}$$

$$\frac{1}{1 + F \times \text{premium}} = \text{SCF}$$

so that Squire and van der Tak [1975] note,

$$\text{SER} = \frac{\text{OER}}{\text{SCF}} \quad \text{and} \quad \text{SCF} = \frac{\text{OER}}{\text{SER}}$$

Border Prices

This term is used in cross-country trading. The price of a traded good at a country's border is known as free on board price [FOB] in case of exports and in the case of imports it is the cost, insurance and freight [CIF] price.

Examples

Problem I

The current prices of shrimp from 1990 to 2000 are given with price index numbers for each year. Convert these prices to constant prices.

Solution

$$\text{Constant price} = \frac{\text{Current price}}{\text{Price index no.}} \times 100$$

Problem II

The cost of labour in a fish farm is Rs. 100 per man-day. Find out the economic cost of labour for skilled and unskilled workers.

Given SCF = 0.75 and 0.8 for unskilled and skilled labourers respectively. (SCF = Standard Conversion Factor).

Solution

$$SCF = \frac{\text{Economic price}}{\text{Financial price}}$$

i.e.,   Economic price = Financial price × SCF

  For unskilled labour

              Economic Price = 100 × 0.75

              Rs. 75 per man-day

  For skilled labour

              Economic price = 10 × 0.8

                        = Rs. 8.0 per man-day.

Problem III

  A. The Fx premium for one dollar is 20 per cent. Calculate the SCF.

Solution

$$SCF = \frac{1}{1 + F \times \text{premium}}$$

$$= \frac{1}{1 + 0.2}$$

$$= 0.83$$

  B. The OER 1 $ is Rs. 45. The foreign exchange premium, for 1 $ is 20 per cent. Calculate SER.

Solution

              ESR = OER [1 + Fx premium]

                  = 45 [1 + 0.2]

                  = Rs. 45 per US $

  C. Given SCF = 0.83 and SER = Rs. 54 per US $. Calculate OER.

Solution

$$SER = \frac{\text{OER}}{\text{SCF}}$$

i.e.,   OER = SER × SCF

                  = 54 × 0.83

                  = Rs. 45 per US $

Problem IV

  A processing firm in India exports shrimp to USA. Here the whole prices are only 4 or explanation, it can be lowered or high.

  Given the cost of production of 1 kg shrimp is Rs. 350;

❑ Export taxes Rs. 20.

❑ Subsidies Rs. 10.

❑ Local port charges including, storage, loading etc. Rs. 5.

❑ Local marketing and transporting costs till port of exporting country Rs. 5.

❑ Freight charges to point import Rs. 20.

❑ Insurance charges Rs. 15.

❑ Unloading from single pier to port Rs. 3.

❑ OER = Rs. 45 per US $, SER = Rs. 54.

Calculate FOB and CIF at financial and economic prices.

Solution

At financial prices,

$$FOB = \text{Cost of production}$$
$$+ \text{ Taxes}$$
$$- \text{ Subsidies}$$
$$+ \text{ Local port charges}$$
$$+ \text{ Local marketing and transporting costs}$$
$$= 350 + 20 - 10 + 5 + 5$$
$$= \text{Rs. } 370$$

$$CIF = \text{FOB} + \text{freight charges} + \text{insurance} + \text{unloading}$$
$$= 370 + 20 + 15 + 3$$
$$= \text{Rs. } 408$$

At official exchange rate

$$FOB = \text{Rs. } 370/45$$
$$= 8.33 \text{ US\$}$$

$$CIF = \text{Rs. } 408/54$$
$$\qquad 9.06 \text{ US \$ (Dollars fluctuates so take consider the recent value).}$$

In economic analysis

$$FOB = [\text{Financial FOB} - \text{taxes} + \text{subsidies}]/\text{SER}$$
$$= [370 - 20 + 10]/ 54$$
$$= 6.66 \text{ US \$}$$

$$CIF = [\text{Financial CIF} - \text{taxes} + \text{subsidies}]/\text{SER}$$
$$= [408 - 20 + 10]/54$$
$$= 7.37 \text{ US \$}$$

Chapter 14

THE ENTREPRENEURSHIP

The capacity of a country for economic growth and development is determined by three key factors : human, physical and financial. Of these human resource factor appears to be, in the final analysis, the most strategic and critical in the absence of which even an abundance of natural and physical resources, machinery and capital, may go grossly underutilized or misused. Thus one of the major tasks confronting the developing countries is the building of human capital formation is as important a precondition of economic growth as is the rapid rate of physical and capital formulation. The human capital would operationally mean here the 'ENTREPRENEUR'.

The growing awareness of the need for and urgency of building entrepreneurs could be ascribed to two factors, First, the *belief is gaining ground* that *economic growth* in the advanced countries appears to be attributable to entrepreneurial awareness in the community rather than to capital. In fact several empirical studies show that the *entrepreneurs as the human capital* has grown in the societies at a much faster rate than conventional capital and has made a larger contribution to economic growth than non-human capital. Secondly, investment in the human *resources* has directly contributed to economic development and growth by promoting the knowledge and application of science and technology to the production process: developing innovations and research: training the workers in different technical skills needed for modern production and building up of the right kind of attitude, values and interests conducive to higher output.

The fundamental problem in developing countries is, therefore, not so much on creation of wealth but the creation of the 'capacity' to create wealth and strengthening, widening and improving the absorptive capacity of the country. Finally, it would not be possible for a country like ours to distribute all needed physical resources in terms of inputs, finance etc., to all those who desire to participate in economic activity. To develop entrepreneurial activity among the persons may create a situation where people become capable of optimal utilization of the limited and scattered resources.

To summarize it again, it may be realized that human beings, like plants and machinery and other physical assets, are important instruments of production; that investment on them is productive; and income yielding as on physical assets and inventories. Therefore, creating entrepreneurial awareness in the society becomes the foremost task.

Entrepreneurial Awareness

A society of community can be said to posses entrepreneurial awareness when we find most of its members are conscious of the important that the entrepreneurs play in accelerating the growth of the economy and in enriching the quality of life in the society. When it is generally realized that man is not necessarily the creature of his environment, he is potentially equipped to be the creator; that change not by what people desire but by what they endeavour to do and are concerned with. The entrepreneurial awareness of the community must lead to detect and appreciate the personality characteristics of entrepreneurs it must help by action the rise of entrepreneurial spirit and in raising the level of entrepreneurial endeavour. It must lead to community leaders and community institutions to participate in the process of prolific emergence and growth of genuine entrepreneurs.

Who is an Entrepreneur?

It should not be astonishing the entrepreneurs have been studied as individuals and groups, to a rather extensive degree. Psychologists are interested in finding out what motivates them. Investors interested in finding out some saliva test that will tell them in advance, which entrepreneurs are going to make them rich. Business schools professors began to be interested in the question of whether and how entrepreneurial behaviour patterns and skills can be taught. Legions of people in governments have begun seeking ways to engage entrepreneurs in programmes of regional development, minority enterprise, leadership training, export promotion and technological transfer, all in the name of common-meal. More specifically in the nature of economic development of the country, particularly in the sphere of industries, the word 'entrepreneur' has been used widely. Let us see the historical the historical background of the terms looking into the specific meaning that different people have attached to this simple word.

The term 'entrepreneur' was coined and applied to business by Richard Cantillon in 1775 an Irishman living in France. According to him the entrepreneur is one who buys factor services at 'certain' price with a view to selling their product at "uncertain" prices in the future (Kilby, 1971, p. 2). Thus the entrepreneur was defined by a unique constructive function: the bearing of non-insurable risk. A few decades later Jean Baptiste Say in 1803 described the entrepreneurial function in broad terms, emphasizing the bringing together of the factor of production and provision of continuing management as well as risk bearing.

Schumpeter's work beginning in 1911 is obvious exception to the above stated ideas. His innovation represents not only the first dynamic concept of the entrepreneurial function but he is the first major writer to put the human agent at the centre.

The recent past literature on defining the entrepreneur's has some distinctive features as compared to earlier concept. Some of the definitions and explanations of the term are listed below:

"Entrepreneur is a highly, respected word in the developed world, it conjures up vision of active, purposeful man and women accomplishing a wide variety of significant deeds. The entrepreneur is an important change in the every society. Yet, he is one of the most enigmatic characters particularly in the less developed world. Although it is his purposive activity that bridges the gap between plan and reality the precise way that this change agent-entrepreneur acts is often unclear".

"The enterpreneur as a dynamic agent of change, is the catalyst who transforms increasing physical, natural and human resources in a corresponding production possibility".

(Schumpeter, 1961)

Entrepreneurship : Different Connotations

Over a period of time the concept of entrepreneurship is enriched by its admirers, which led towards manifestation of its characteristics from different angles. The important factors, which influence its development according to the various researchers, are summarized below.

Summary of finding of various researchers

Author Factors that contribute to development of entrepreneurship	
Schumpeter, 1961	Suitable **intuition** in grasping the essential facts.
Weber 1961	**"Protestant Ethic"** which emerged form the religious belief system and which is absent in oriental religious belief system.
Levin 1969	**Status mobility**, system where status is attained through standing performance, **initiative, industriousness** and **foresight** through **self-reliance** and **training.**
Hagen 1971	**Creative** personality, high **needs achievement, need order** and **need autonomy.** Fairly widespread creative **problem solving ability,** and a tendency to use it. **Positive attitudes** towards manual and technical labour, and the physical world.
Cochran 1971 McClelland 1969	**Positive** attitude towards occupation, the **role expectations** held by sanctioning ground, and the operational requirement of the job. Need for achievement thorough self-study **goal setting,** and inter-parson support. Keen interest in situation involving **moderate risk,** desire for taking **personal responsibility** concrete **measures of task performance, anticipation of future possibilities organizational skills** and **energetic.**
Kilby 1971	**Perception of market opportunities,** gaining **command over scarce resources,** and **marketing of products.** Dealing with public bureaucratic concessions, licenses, taxes management of **human relations** within the **firm and with customers** and suppliers and production management technological knowledge.
Nafziger	Perceived **challenge to status, migrants, new religious sects,** and 1971 **reformed groups.**
Staley and Morse 1965	Quality of service in industrial advice, managerial training and industrial research.
Fox 1973, Mines 1973, Papanek 1973	Economics opportunities and political conditions.
Nandy 1973	**Supportive community,** self-image that gives meaning values, and status to an entrepreneurial career.
Singer 1973	Traditional system of occupational culture which facilities the process of modernization, special opportunities, motivations, experience training or knowledge. Traditional belief and value system, which are flexible to allow for reinterpretation with changing conditions.

IMPORTANT PERCEPTIONS

Sociologists View

Early sociologists (like Max Weber, 1930) suggested that the belief systems of Hinduism, Buddhism, and Islam, did not encourage entrepreneurship. This contention has, however, been challenged and refuted by many sociologists like Fox (1973), Mines (1973), Papanek (1973), Nandy (1973) and Singer (1973).

Psychologists View

Psychologists view achievement motivation aspects, and linked these with the nature of socializations in the society. In this, there was overemphasis on the individual and his values, attitudes and personality. This however has been criticized by as Kilby (1971) and Kunkel (1971).

Economists View

The economists assume that the factors of production, mobility of inputs and outputs, and that producers, consumers and resource owners have knowledge of all the possibilities open to them. In an underdeveloped country such ideal conditions do not exist.

Managerial View

Managers emphasized perception of market opportunities as well as operational skills, required to run a business or an industry.

These four thoughts, however, brings to surface certain common characteristics. The include the perception of economic opportunity, technical and organizational skill, managerial competence, and motivation to achieve results.

Indian Experience

A number of western social scientist also supported Weber's theory at that time. The powers readily accepted and encouraged such thinking. It provided them with a convenient rationale for ignoring and even discouraging the industrial development of the colonies. In the pre-independence days in India the British rulers tried to perpetuate thinking that whatever industrial development took place in India at that time, was because of their presence, and that such development would wither away if they left.

These often ignored historical facts underline the capacity of India society for adaptability, flexibility and tolerance to new ideas. Indian religious system, as has been made out by some researches, therefore, need not stand in the way of member of the society becoming entrepreneurs.

Indians and people of other developing countries have in the business of goods and services before the developing countries had started. With the development of science and technology which the developing countries picked up faster, entrepreneurship touched a new height. What changes after independence took place in industry was also taking place in the agricultural sector. Indian farmer has shown no hesitation in adopting modern agricultural technology.

In India, today, the term entrepreneurship appears to connote a much-restricted meaning. It covers only a limited sphere of enterprising endeavor. Here the thinking generally veers round effort, which result in establishing and running factories and industrial enterprises alone. Moreover, there is a marked tendency to relate it only to operations, which exceed a particular size. This narrow overview of the concept, perhaps, reflects the preponderance of values nurtured by urban white-collar class in this society. Secondly, entrepreneurship has been viewed as a phenomenon occurring around an individual and benefiting only an individual. It is rarely appreciated as one that could be harnessed to benefit larger groups. Like wise there is a marked tendency to view its occurrence only in terms of the total aggregate society. This approach ignores the distinctly varying social environments which confront the large variety of smaller groups, and which present both opportunities as well as challenges dissimilar in nature. The idolatry consideration of entrepreneurship has led to inadequate appreciation of the role that the joint family system has played in this country in the sphere of economic development. It is often ignored that industrialization requires changes not only in the modes of production resulting form the use of modern technologies and machines, but also other spheres of associated activities like trade and commerce, more particularly in the ways of managing production and commercial activity. Secondly, it is often not adequately recognized that the process of transformation form the rural and agricultural society to industrial society, would have to cover all sizes, shapes and types of economic activities.

In fact in India society today, use of new techniques and equipments on an universal scale alone holds adequate promise of industrialization. The use of modern techniques and equipment in a small wayside repair shop or machines like the tractor in place of bullock-drawn ploughs on farms, does contribute to processes of industrialization as much as the use of sophisticated, complex techniques and processes in a modern electronic factory. In all such cases the person using modern devices an equipment could be considered an entrepreneur, provided he is innovative, he is influenced by the urge to take risks, and has been acting on the spur of the intuition which proves to be right afterwards. The belief that even an illiterate or semi-literate person could also play entrepreneurial roles, does not seem to have covered enough ground as yet. Perhaps because of the lingering influence or values nurtured by the urban elite.

It is more often not realized that in a stratified society like the one in India, the indicators that denote entrepreneurship may not be uniform for its different strata. What could be entrepreneurial activities in a particular of population may not be considered so in other strata. When a tribal person, for example, opens a wayside shop and opts business, it is an entrepreneurial event in his immediate social circle. While the same exercise by a person belonging to a trader community would not be treated as such, in the social environment of the latter. Our definition, therefore, have to be adequately perceptive of such diverse social phenomena. Seen in this perspective, occupational mobility would be one of the key indicators of entrepreneurship in developing societies.

Similarly it is not recognized that the entrepreneur is not always motivated by self-interest alone. He could also have collective, corporate interests, which guide him into action. Leadership in cooperative such as cooperative sugar factories appears to have from such motivations.

Indian society has a composite population. It consists of groups belonging to different religious and castes, and in certain pockets even to different tribes. Simultaneously it is stratified. The caste system determines the socio-religious, political and economics status in the society. Traditionally caste is an endogamous group. Its members follow certain occupation. Over centuries each caste has evolved its own way of life, food habits, socio-religious customs and rituals. Each caste is further subdivided into numerous subcastes, which have again developed certain common characteristics that give them separate identities. In recent year spread of education, has created new job opportunities. In addition, the process of urbanization and industrialisation has tended to divest the castes of their occupational identities. These new influences are largely restricted through to urban and industrial centers leaving the bulk of the rural population out of its purview.

Like desirability of simultaneous technological change at all levels of productive activity, the flow of entrepreneurs form all strata of the population also may be considered a prerequisite for accelerated economic development in the Indian context. Creativity need not be considered a monopoly of any particular strata of the population or a caste. The potential for creativity in one field or the other always existed in all strata. In the past, the socio-economic environment was not conducive to adequate expression to the potential. The simmering dissatisfaction with the traditionally imposed social status among the lower strata or society is indicative of this urge to seek new expression. With the increasing opportunities for education, which hold a promise of occupational mobility and the diversity of job-opportunities thrown open by the processing of industrialization and urbanization, a new social-economic environment conducive to entrepreneurship appears to be emerging fast.

Such motivation to break with the past (even if it meant a little risk), and the widening perceptions of economic opportunities may be considered as two key factors holding out a great promise for enterpreneurial development. These key factors may have to be supported by organizational and managerial skills and competence. Together they hold a promise of ever expanding entrepreneurial possibilities.

REFERENCE

Extracted from the article of V.R. Gaikwad, Entrepreneurship : The Concept and Social Context, (Ed.) T.V. Rao and Udai Pareek Developing Entrepreneurship : A Hand Book, 1978.

RELATIONSHIP BETWEEN PROFILE CHARACTERISTICS AND KNOWLEDGE GAINED

Indian Agriculture aquaculture has made impressive strides in the development of new plant varieters, cultivars; hybrids and production and standardization of plant protection techniques.

However, quick dissemination of technological information from the Agricultural Fisheries Research System to the farmers in the field and reporting of farmers' feedback to the research system is one of the critical inputs in the Transfer of Agricultural allied/fisheries technology (Sharma, 2003). Farmers can no longer depend on the conventional and time consuming manual dissemination of technological messages.

To reach over 110 million farmers, spread over 500 districts and over 600 block is an up hill task. The diversity of agro-ecological situations adds to this challenges further. Farmers needs are much more diversified and the knowledge required to address them is beyond the capacity of the grass root level extension functionaries.

It is in this context that artificial intelligence based computer programmes called Expert System receive a great deal of attention by virtue of its dynamic, heuristic strategies and ensure a speedier and more effective transfer of farm technologies.

Keeping these ideas in view, a study was undertaken with the following objectives.

1. To study the knowledge gain due to exposure to the various treatments.
2. To study relationship and influence of profile characteristics of the subject (rubber growers) with the knowledge gain using the various treatments.

The knowledge of bioin for matics leads to predict abiochemice like protein structure, biotechnological and gene tools sequences etc.

A computer based Expert System for rubber protection technologies (RUBEXS-04) was developed using knowledge engineering and software engineering components. Multiple group randomized design was used to establish adequate relationship (independent variable) between the profile characteristics and knowledge gain (Dependent variables).

The service area of Rubber Board Regional Office, Mannarkkad, Kerala state was selected for the study. Out of the 60 existing rubber producers societies in this region, three societies were randomly selected. A sample of 40 rubber growers from each society was drawn randomly. Thus a sample of 120 rubber growers formed the total sample for the study.

Four different treatments such as human experts without discussion, human experts with discussion, RUBEXS-04 without discussion and RUBEXS-04 with discussion was selected by the researchers. These Treatments were tested for their relative effectiveness. Each treatment was replicated thrice. Considering 10 respondents per replication there were 30 respondents per treatment. Thus a total of 120 respondents were the subjects for the four treatments.

'Before-After' techniques of measurement was used to find out the effect of a particular treatment.

Suitable statistical techniques such as 't' test, correlation analysis and multiple regression analysis were used to analyse the data.

Table 1 : Mean knowledge gain due to exposures to the treatments

Treatments	Mean knowledge score		Mean knowledge gain	Per cent of knowledge gain	't' value
	Before exposure	Immediately after exposure			
Human expert without discussion (T_k1)	6.86	9.60	2.74	13.04	-6.40**
Human expert with discussion (T_k2)	7.26	11.23	3.97	18.90	-6.075**
RUBEXS-04 without discussion (T_k3)	6.33	9.00	2.67	12.71	-3.387**
RUBEXS-04 without discussion (T_k4)	8.43`	15.23	6.80	32.38	-9.944**

**Significant at 0.01 level

It could be observed from Table 1 that all the four selected treatments namely, human expert without discussion (T_k1), human expert with discussion (T_k2), RUBEXS-04 without discussion (T_k3) and RUBEXS-04 with discussion (T_k4) had highly significant 't' value indicating that all the four treatments were effective in terms of knowledge gain.

The mean knowledge gain was maximum with a score of 6.80 in RUBEXS-04 with discussion which indicated 32.38 per cent of knowledge gain. This was followed by human expert with discussion with a score (2.74) and RUBEXS-04 without discussion (2.67) which resulted 18.90 per cent, 13.04 per cent and 12.71 per cent knowledge gain respectively.

The above results clearly indicate that all the four selected treatments were effective in imparting knowledge on plant protection aspects of rubber crop with considerable variation in their effectiveness.

Relationship and influence of independent variables towards knowledge gain

The relationship of the 16 independent variables with the dependent variable knowledge gain is shown in Table 2. As could be seen from the Table, of the 15 independent variables studies, (Independent variable 10, had a missing correlation and hence was excluded in the analysis) only three namely, area under rubber cultivation, experience in rubber cultivation and information seeking behaviour were found to be positive and having highly significant relationship with knowledge gain at one per cent level of probability. Whereas the other three variables namely age, possession of modern electronic gadgets and familiarity in using computer exhibited a positive and significant relationship at five per cent level of probability.

Table 2 : Relationship of independent variables with knowledge gain

Variables	'r' value
Age	0.202*
Education status	0.073[NS]
Occupational status	0.103[NS]
Area under rubber cultivation	0.300**
Experience in rubber cultivation	0.306**

Contd...

Variables	'r' value
Annual income	0.061[NS]
Communication status	0.033[NS]
Information seeking behaviour	0.250**
Possession of modern electronic gadgets	0.184*
Training undergone on computer	@
Familiarity in using computer	0.218*

"An entrepreneur is one who is willing and able to initiate and successfully manage for a length of time an activity that involves at least some degree of personal and organisational risk".   *(Khandwalla, 1979)*

"The person using modern devices and equipment could be considered as an entrepreneur, provided he is innovative, he is influenced by the urge to take risk and been acting on the spur of intuition which proves to be right afterwards".   *(Gaikwad, 1978)*

"An entrepreneur is one who initiates and establishes an economic activity or enterprise".   *(Pareek & Nadkarni, 1978)*

"An entrepreneur is a person having characteristics such as need for achievement; need for influencing others, sense of efficacy, risk taking, openness to feedback, need for independence, hope of success, a belief that he can change the environment, Time orientation, concern for society, self consciousness dignity of labour, competition and collaboration, saving for future etc. More these characteristics are present in a person the more effective he is likely to be as an entrepreneur".   *(Rao & Mehta, 1978)*

On the basis of definitions given by different authors it may be stated that the entrepreneurs is perceived as an individual with certain characteristics helpful in conceiving, initiating, establishing, running and finally managing an enterprise. An enterprise can vary form starting a small shop to establishing an advanced technology based industry. *An entrepreneur therefore, may be differentiated not only in terms of the kind of activities he pursues but in the context of his life style,* attitudes, values and behaviour which together go to make the entrepreneurial personality.

The Entrepreneur : Some Prevalent Myths

Despite the fact that the entrepreneur has been defined and redefined by historians, economists, sociologists, psychologists and behavioural scientists, we find some misconceptions arising on the mind of those who deal with entrepreneurs in the developmental process. Examining the misconception/myths in one way of clarifying our concepts in this regard.

The Entrepreneur's Primary Motivation is a Desire for Wealth

Perhaps the most misunderstood aspect of the entrepreneur is his or her relationship to money. Popular opinion generally holds that entrepreneurs are driven by greed, that fundamental to their character is just for money that them to do things which ordinary people would not do. The fact is that money is very rarely the primary driving force for successful entrepreneurs (Gifford Pinchot, 1985). Their attitude towards money is complex and intimate. They do care about it and work for it but is not the thief goal in their life. As Hallmark Cards' founder J.C. Hall (1979) put it. "It a man goes into business with only the idea of making a lot of money, the chances are won't."

What drives the entrepreneurs is a deep personal need for achievement but that need generally becomes wedded to rather a specific vision of what he wants to accomplish. As Mc Clelland (1965) stated:

"He (the entrepreneur) does not seem to be galvanized into activity by the prospect of profit, it is people with low achievement need who require money incentives to make them work harder. The person with high need for achievement works hard anyway, provided there is an opportunity of achieving something. He is interested in money rewards or profits primarily because of the feedback they give him as to how well he is doing. Money is not the incentive to effort but rather the measure of is success for the real entrepreneur".

Entrepreneurs are High Risk Takers

Popular belief holds the entrepreneur as a daring, devil-may-care risk taker. The common saying- "no risk-on gain" is often viewed as implying that a very high order of risk is required to establish an enterprise where fortune and chance play a vital role. The term 'risk' commonly refers to as outcome which leads to losses or deviations of realizations from expectation. This simple meaning of risk, however, does not seem to be applicable in context of entrepreneurial behaviour. Risk taking willingness in case of an enterpreneur indicates a challenge in his activity where there is reasonable chance of success. Success depends not on chance but on one's own efforts. McClelland (1985) argued "On of the striking characteristics of entrepreneurs is their willingness to take calculated risk to innovate in ways that have reasonable change of success". Studies show that successful entrepreneurs avoid high risk situations, rather they seek and enjoy calculated moderate risk. They do choose challenging goals, but they also do everything they can to reduce the risk.

Part of the entrepreneur's strategy for reducing risk is anticipating barriers and remaining open to feedback, both positive and negative. One who can not see problems or imagine how anything might go wrong may be more aptly called a 'promoter', not a real entrepreneur.

Entrepreneurs are Amoral

Perhaps the most striking similarity of all venture capitalists' descriptions of the entrepreneur is their insistence that honesty and integrity are characteristics of the successful entrepreneur. Some may find this surprising because in the popular mind entrepreneurs are often seen as willing to sacrifice morals for profit. But it is less surprising when one considers that entrepreneurs are generally deeply committed to what they consider to be worthwhile purpose. Their need to achieve produces flexibility with the rules, not a loss of integrity.

One venture capitalist described his idea of successful entrepreneurs like this "they are darned honest with themselves. It there is a problem, they tend to get it out in the open fast and then stick with in until is solved". Another said of his entrepreneurs "All are extremely honest with themselves and will not tolerate untruthfulness or dishonesty."

Since entrepreneurs often have to handle dozens of functions that they know little about, this ability to steer through truthfully is essential.

Entrepreneurs are Power Hungry Empire Builders

Watching entrepreneurs build large organisations with themselves at the helm. It is easy to imagine that they are driven by the need to tell others what to do. But it turns out that the need for power is not an important part of the entrepreneurial motivation. Power-driven people are satisfied to achieve things by getting others to do them. However, we may find it quite reverse in case of entrepreneurs.

Lying behind motivation, McClelland felt, are the fantasies every person has of what he or she wants to be or do. Based on his extensive studies of entrepreneurs he concludes that entrepreneurs are not driven by a need for power, instead, their motivation stems form a very high need for achievement. Entrepreneurs, he found, are not so concerned with the corner offices, large number of people to tell what to do and imposing possessions. They are not satisfied with rising in the hierarchy and having the esteem of their peers. Instead, entrepreneurs are driven by the need to achieve - to leave behind their mark by accomplishing things that have never been done before. Therefore, the myth of the power-motivated entrepreneurs turns out to be false. Entrepreneurs are less power-motivated than executives and the fact that they are achievement oriented instead explains both their strengths as business starters and their potential weakness as executives.

Entrepreneurs are Inventors

It is commonly believe that an entrepreneur is basically an intelligent person and has a definite ability to create something new to prove his worthiness. On the basis of the existing profile of a successful entrepreneurs it has been found clearly that entrepreneurs are innovators, not inventors.

Invention and innovation are two entirely different aspects. Invention provides the initial insight or discovery that as scientific or technological problem can be solved, whereas 'innovation' is the socio-managerial process by which new products and techniques are introduced into a socio-economic system. It has been established that not all inventors are good innovators. Henry Ford is a good example. He did not invent a thing. He drew upon the ideas of others and put them together on the assembly line of his automobile industry. There was nothing new in these techniques. But Ford was an innovator and entrepreneur. On the other hand, Thomas Edison was an inventor. He went bankrupt once, and though he survived the disaster, he could not make money in his life time. He was not an innovator and entrepreneur. To give yet another example, Chester Carlson, the physicist, invented a document-copying device in 1930, based on his knowledge of photo conductivity, carrying out his experiment in this own kitchen. But it was only in 1960 that a creative group under Joseph C. Wilson of the Haloid Corporation exploited the invention fully to bring about a major revolution in the field of giving us xerox copying. Thus enterprises require not only brilliant ideas but also creative ideas harnessed to a productive drive in order to work out the invention into a product.

Entrepreneurs are Born : They are Built

A common belief that the entrepreneurs are born has been another block in understanding fully the key element in developing entrepreneurship. Emergence of entrepreneurs in few restricted castes and regions of people more often then others, perhaps, made the belief more stronger even among those promoting entrepreneurship. People belonging to some of the castes like Marwari, Sindhi, Punjabi etc. Our generally believed to have in born entrepreneurship qualities viewed objectively, it is not difficult to accept that those who take birth in these castes are definitely in advantageous position as they get an opportunity to be aware about business environment more than those from other castes. However, it has been clearly established that emergence of entrepreneurs is independent of caste/region. Any one with certain entrepreneurial characteristics at least to a minimum level qualifies to be an entrepreneurs.

Some of the important characteristics of an entrepreneur are psychological in nature. These characteristics can be influenced through training. The kind of influencing that is attempted in training perhaps, also takes places as the process of socialization of the child & the youth in the family and the community for the traditional entrepreneurial caste groups. In other words, the characteristics in an individual are heightened to a required level which finally give rise to a unique kind of manifestation that may be termed as entrepreneurial behaviour. This is exhibited while conceiving and establishing the enterprise. Some people have gone far in over simplifying by saying that "Entrepreneurs can be manufactured" which no doubt does equal disservice to the cause of entrepreneurship development.

The misconceptions stated earlier seem to have define implications in promoting entrepreneurship in any society. Since the entrepreneur are the key factor in developing entrepreneurship, any misunderstanding about this key element may create a basic problem of integrating other aspects like support facilities with them. Even worse, problem may arise in terms of treating the entrepreneurs in the right spirit. It is established fact that the entrepreneurs do seek positive recognition in the society and therefore, would not like to be treated like any venture capitalists or businessmen. To preserve the dignity and prestige it appears to be important that all those engaged in promoting entrepreneurship understand the entrepreneurs as what they are rather than what they are not.

Every entrepreneur may not have all the characteristics. In fact there is not much research to indicate this. However, it may be stated that the more these characteristics are present in a person the more effectiveness likely to be as an enterpreneur. The person should be devoted to achievement motivation we have to see Need for Achievement, Need for influencing others, Sense of Efficacy, risk taking openness, need for independence, Hope of Success, A belief in adversity, time orientation.

The need to excel, known as achievement, is the single psychological factor that has been extensively researched in relation to entrepreneurship it has been demonstrated that achievement motive is a critical factor for entrepreneurship. Contribution of achievement motivation is basic in helping people become entrepreneurs. Through training entrepreneurs by and large, have been found to be people with a high drive and high activity level constantly struggling to achieve something, which they could call as their own accomplishment. They like to be different form others and strive to accomplish goals. At the same time they do not strive to achieve something, which is practically impossible to accomplish.

Achievement orientation can be induced by enriching their thinking and fantasy world with achievement language. Many organizations round the world have started giving achievement motivation training programmes. In development entrepreneurs have made such training an integral part of their inputs. In India the Small Industry Extension Training Institute at Hyderabad started giving national and international Experiments indicate that achievement orientation can be administered. Many organizations have courses on achievement motivation. Behavioral Science Center (India) commenced training of trainers for such courses. Now-a-days it is quite popular (This activity has since been taken over by National institute of Motivational and Institutional Development) Amongst government-sponsored organizations, the Gujarat Industrial investment Corporation has been the pioneer in giving the achievements motivation training for its entrepreneurial trainees.

Small Scale Industries Development Corporation, Indian Institute of Technology, Delhi and several other organizations are also providing achievement motivation training. The results of these training efforts have been quite encouraging.

In developing these entrepreneurial characteristics we examined the available research and theory about the behavioural and other characteristics of successful entrepreneurs and new ventures. We also researched the current practices of venture capitalists in assessing the personal characteristics of entrepreneurs in whom they invested. We found there was considerable effort to determine if persons wanting to start a new venture, and applicants for venture capital did, in fact, possess the behavioural and other characteristics of successful entrepreneur. Our investigations identified fourteen dominant characteristics of successful entrepreneurs.

1. Drive and Energy

Entrepreneurs have a tremendous amount of personal energy and drive. They possess a capacity to work for long hours and in spurts of several days with less than a normal amount of sleep. Our research in the venture capital industry confirmed drive and energy as a characteristic desire by investors and very frequently observed in successful entrepreneurs.

2. Self-Confidence

There is also agreement that successful entrepreneur have a high level of self-confidence. They tend to believe strongly in themselves and their abilities to achieve the goals they set. They also believe that events in their lives are mainly self-determined and they have a major influence on their personal destinies, and have little belief in fate. Venture capitalists and the financial community also look for a strong sense of self-confidence in the entrepreneurs in whom they place their money. But they are negative about over-confidence. Subtle arrogance or lack of humility which may suggest a lack of realism.

3. Long Term Involvement

This is one of the characteristics which distinguishes the entrepreneur — the creator and builder of a business — from the promoter or fast-buck artist. Entrepreneurs who create high potential ventures are driven to build a business, rather than simply get in and out in a hurry with someone else's money. They make a commitment to a long-term project and to working towards goals that may be quite distant in the future.

One venture capitalist has a saying about how long it takes to build a business: the lemons ripen in 2 or 3 years; the plums take 7 or 8 years.

4. Total Immersion and Commitment

Launching and building a business in note a part-time proposition. And *building a high-potential* business requires a total immersion in and *commitment to that end*. An earlier researcher noted that the managers role in business could be delegated, while the entrepreneurial role could not Today's venture capitalists are in agreement with this view. Ned Heizer of the Heizer Corporation, a $ 80 million venture capital firm has said 'You can hire managers, they're a dime a dozen, but you can't hire an entrepreneur.

The commitment requirement for high-potential entrepreneurship is measured in part by some of the expectations the venture capitalist has for gauging *commitment; a willingness to invest your life savings,* a substantial portion of your net worth, to reduce your income by as much as one-half for the first year or two, to second mortgage you home, not to mention long hours and neglected family left. In addition, you may very well be asked to execute an investment agreement which states that you lose every dime of your own the venture before the investor loses a single penny.

Building a new business in a way of life, and unless it is a way of life you thrive on, it can be intolerable. Perhaps Joseph Conrad captured the meaning of this king of work for the entrepreneur although unintentionally.

"I don't like work—no man does—but I like what is in work—the chance to find yourself. Your own reality—for yourself not for others—what no other man can ever know.

5. Creativity and Innovation

The entrepreneurial role has long been recognized as a prime source of innovation creativity. More recent research has shown career patterns of entrepreneurs follow paths that afford opportunities to be creative and innovative. Recent studies of entrepreneurial careers have identified different career anchors' associated with different roles. Once again, the entrepreneur's career anchor is creativity and innovation. *Managers, on the other hand, find their career anchor in competence and efficiency,* while college professors prefer autonomy having control and discretions over one's time—as a career anchor.

6. Knowledge of the Business

Venture capitalists stress the importance of the track—the business experience and accomplishments—of the entrepreneur and team they invest in. Most strongly prefer that a *potential venture be headed* by a *entrepreneur who has* through and *proven operating knowledge of the busines* they intend to launch. Hindsight has made *even more venture capital* investor *wish they had* insisted upon this requirement and *several* studies support this view. Formal-education has not been a distinguishing-requirement for successful entrepreneurship and some believe that more education versus less *can be liability. But when it comes* to technical and scientific based entrepreneurship formal education related to the kind of business one intends to start does appear to be a requirement.

People and Team Building

High-potential ventures are rarely sole-proprietorship. Venture capitalists place considerable emphasis on the demonstrated capacity of the lead entrepreneur to attract, motivate and build a high quality entrepreneurial team. Aspiring entrepreneurs who are adverse to having partners must recognize the significant implications of a decision to go it alone. First, they will be generally unable to raise venture capital form conventional sources. They will have to rely on sweat equity, personal net worth, friends and family for seed capital. Second, their business, with rare exceptions, will not be able to grow much beyond the $ 1 million annual sales level without a supporting management team.

Economic Values

Business (unlike social or non-profit) entrepreneurs must share the key values of the free enterprise system; private ownership, profits capital gains responsible growth. These dominant economic values need not necessarily exclude social or other values. But the realities of the competitive market economy seem to require a belief in or at least a tolerance of these values.

Personal values can have a profound effect on the development of the ream and the business itself. An exception to this generalization appears to be in the area of high technology entrepreneurship. Komives in his study of high technology entrepreneurs found that aesthetic and theoretical values were strongest.

Ethics

In this post-Watergate era the ethics issue is delicate and controversial. It is generally conceded that the business community has its own ethical standards and that they probably differ markedly from the religious community for instance. Historically, the entrepreneur has tended to possess what we refer to as situational ethics. Personal ethics of the entrepreneur tend to be defined by the needs and demands of the situation rather than by some external rigid code of conduct applied uniformly regardless or different conditions and circumstances.

Integrity and Reliability

Not to be confused with situational ethics integrity and reliability are a must for an aspiring entrepreneur to raise conventional venture capital and debt financing. An early eliminator of a venture proposal is the discovery by a venture capitalist that the entrepreneur has not been straightforward and honest own experience venture capital investors know that the successful building of a growth business requires a total immersion and concentration on the attainment of the distant goal. This long-term involvement that is characteristic of successful entrepreneurs has also been confirmed by research.

Money as a Measure

Money has a very special meaning to the successful entrepreneur. It is a tool and a way of *keeping score. Profits, capital gains and net worth are seen as* measures of how well the *entrepreneur is doing in pursuit of self established goals. Many* examples exist in the venture capital industry to illustrate this notion. One entrepreneur, for example, successfully built up his business and *sold it for several million dollars.*

Persistent Problem Solving

Entrepreneurs who successfully build new enterprises possess an intense level of determination and desire to overcome hurdles, solve a problem and *complete the job.* They are not intimidated by difficult situations. In fact *their self-confidence* and general optimism seems to translate into a views that the impossible just takes a little longer. Interestingly enough, if the task is extremely easy or perceived to be unlovable, the entrepreneur will actually *give up sooner than others.* Other researchers share this view that while entrepreneurs are extremely persistent they are also realistic in recognizing what they can and can not do, and where they can get help to solve a very difficult but necessary task.

Goal-Setting

Entrepreneurs are very goal-oriented. They have an ability and commitment to set clear goals for themselves. These goals tend be high and challenging but they are achievement. Related to this in the short-term is a great concern for time. They will tend to set their watches a few minutes fast in order to avoid lateness. They are appalled at wasted time. Hence, having clear, measurable goals is an effective way for entrepreneur to set priorities, measure and guide their time allocations.

Personal accomplishments as well as *setbacks lie within one's personal control* and influence. This sense of personal causation as the determinant of successful failure is linked to the entrepreneur's achievement motivation and preference for moderate risk lacking. Several research have reported a positive correlation between one's entrepreneurial activity and the entrepreneur's belief that the locus of control over these entrepreneurial events is internal rather than external and just a matter of luck or circumstances.

Tolerance of Ambiguity and Uncertainty

Entrepreneurs have long been viewed as having a special tolerance for ambiguous situations and for making decisions under conditions of uncertainty. It contrast to the professional manager, entrepreneurs are able to live with modest to high levels of uncertainty concerning lob and career decisions and security. Job security and permanency are considerably lower on the entrepreneur's hierarchy of preferences compared to his/her managerial counterpart.

Role Demands and Requirements Entrepreneurs Face

It is not enough to simply posses a large and intense level of these entreprenurial characteristics. In addition, certain conditions, pressures and demands are inherent in the role of entrepreneurship. These role requirements have important implications for one's fit with the entrepreneurial task, and for the eventual success or failure of a venture. *While successful entrepreneurs may share several characteristics in common with successful persons in other careers, their preference for and tolerance of the* combination of requirements unique to the entrepreneurial role is a major distinguishing feature. Researches identified eight dominant role requirements.

Accommodation to the Venture

The entrepreneur lives under huge, constant pressure first to survive, then to play alive, and always to grow and withstand competitors' thrusts. *The high potential venture demands top priority for the entrepreneur's time, emotions,* and loyalty. Professor Edgar Schein of MIT has studied several graduating classes from the Sloan school in the 1950's. He found a familiar pattern among entrepreneurs and general managers alike. For the entrepreneurs their businesses came first by a substantial margin, as did the careers of general managers. Our own research confirms this. *Entrepreneurs must be prepared to 'give all' to the building of the business, particularly in the early standup years. This demand has important implications for decisions relating to marriage, raising a family and community involvement.* On the other hand research of small business owners whose venture are not in the 15-20% of all ventures which are growth oriented high potential firms shows there is room for accommodating family and community priorities without damaging the business. In fact, some writers indicate that owners of small business are probably dominated as much by personal and family consideration as by the profitability of the business.

Moderate Risk Taking

The successful entrepreneur prefers to take moderate, calculated risks where the chances of winning are neither so small as to be a gamble nor so large as to be a sure thing. Rather, risks are preferred which provide a reasonable and challenging chance of success, and a situation whose outcome is influenced as much by one's ability and effort as by more chance. This entrepreneurial characteristic is one of the most important, since it has such significant implications for the ways decisions are made, and thus for the success or failure of the business.

They seem to be very much aware of the no risk—no return, risk, high return continuum, and to take moderate, challenging risks where moderate returns are attainable and more influenced by their abilities and decisions. Several studies have identified or confirmed the importance of this achievement motivated characteristic

Dealing with Failure

Entrepreneurs are not afraid of failing. Being more intent on succeeding they are not averse to the possibility of failing. The persons who fear failure will neutralize whatever achievement motivation they may possess. They will tend to engage in the very easy task, where there is little chance of failure; or in a very difficult (chance) situation, where they cannot be held personally responsible if they don't succeed.

Use of Feedback

Entrepreneurs, as high achievers, are very concerned about their performance, especially about doing well. This concern is responsible in part for this entrepreneurial characteristic: use of feedback. Without information or feedback about performance the entrepreneur cannot know how well or poorly he is doing. Successful entrepreners demonstrate a capacity to seek and use feedback on their performance in order to take corrective action and to improve.

At an earlier age this tendency shows itself in the preferences children have for sports and hobbies. For instance, entrepreneurial types like competitive sports, and building things with their own hands. These provide excellent opportunities for a sense of personal accomplishment where feedback is readily available. In a similar vein, natural mechanical ability and inclination often are associated with the entrepreneurial type person, and is related, in part, to this characteristic.

Taking Initiative and Seeking Personal Responsibility

The entrepreneur has historically been viewed as an independent and highly self-reliant innovator, the champion (and occasional villain) of the free enterprise economy. Modern research and investigation into the entrepreneurial personality have confirmed some of these earlier generalizations, but have refined considerably ways of focusing on this self reliant aspect. There is considerable agreement among researchers and practitioners alike that effective entrepreneurs actively seek and take initiative. They willingly put themselves in situations where they are personally responsible for the success or failure of the operation. They like to take the initiative to solve a problem or fill a vacuum where no leadership exists. They also like situations where their personal impact on problems can be measured.

Use of Resources

Several studies have emerged in recent years which show that successful entrepreneurs know when and how to seek outside, as well as inside, help in building their companies. Successful entrepreneurs seek expertise and assistance that is needed in the accomplishment of their goals. They are not so ego involved in purely individual achievement of goals and independent accomplishment that they will not seek aid or let anyone help. This characteristic, at first glance, appears to be at adds with a popular stereotype of all entrepreneurs as highly individualistic and self-reliant loners. The willingness to seek and to utilize outside resources is one key characteristic which we feel distinguishes the high-potential entrepreneur.

Competing Against Self-imposed Standards

Competitiveness by itself can be a misleading attribute. It is most important to distinguish between competition directed toward others, without any objective measure of performance, and competition with a self-imposed standard competition directed toward beating the other person (as in boxing) can have in it seeds of self-destruction, for example, price wars below variable costs. This kind of competition has an externalized standard (the opponent) or no objective standard at all. This orientation tends to be more reactive to what others are doing, then proactive in figuring out what to do before the fact based on your goals, capabilities, resources, etc.

Competition with self-imposed standards is an internalized kind of competition. The most obvious example of this kind of competitor is the world class swimmer or runner. They repeatedly indicate that they perceive their opponent as the clock, rather than the other swimmers or runners.

Similarly high performing entrepreneurs also possess this internalized competitive spirit in which she continuously engages in competition with himself/herself to beat their last best performance.

Internal Locus of Control

The entrepreneur does not believe the success or failure of a new business venture depends mostly upon luck or fate, or other external, personally uncontrollable factors. Rather, the entrepreneur tends to believe that one's in revealing past dealings of even failures. A reputation for dependability reliability and honest in a must.

These fourteen entrepreneurial characteristics and eight role requirements have been distilled from the literature on entrepreneurship, our research with venture capitalists and our practical experience. These dimensions comprise what we currently believe are the most important aspects of entrepreneurial behaviour. The biggest gap in existing knowledge is how to measure accurately and consistently the psychologically oriented characteristics. We also hasten to emphasize that we have to see a single entrepreneur who possessed all of these fourteen characteristics to an extremely high degree. In principle, more are better than fewer, but weaknesses can be complemented with other team member's strengths. Thus, we contend it is more important for entrepreneurs, investors and advisors alike to know thoroughly the entrepreneur's strengths and weaknesses with respect to each of these characteristics.

The eight role dimensions comprise what we currently believe are the most important demands of entrepreneurship. Other than the area of personal value we know of no tests or other instruments which can readily asses one's probable fit with a particular role requirement.

Potentially, these criteria have implications on at four fronts. Teachers and counselors can help would be entrepreneurs make a more candid and systematics evaluation of their fit with an entrepreneurial career. Second, other practitioners — venture capital investors, loan officers, personnel — can supplement and cross check their own proven rules-of-thumb with these criteria. Third, would be entrepreneurs, as well as entrepreneurs who are re-evaluating their careers, may find these criteria useful in sorting out capabilities and attributes. Lastly, those interested in researching the entrepreneur should find rich opportunities on numerous fronts. For instance, just how generalizable are these criteria as we very industry, technology, stage of growth or size? How can be develop better empirical evidence to confirm or dispute existing data? And this, in turn implies a major opportunity for research into testing, evaluation techniques and other methods to further refine this knowledge base.

Dealing with people effectively needs a drive which influences people, a drive, which sells them his ideas, and lead them to the process of establishing and expanding his organization. Such drive to influence people and to lead them implement his ideas may be called as need for power. Entrepreneurial managers have been found to have this need high. Such people successful institution builders and have a great need to influence others in their environment and a strong belief in their own capacity to succeed. They can predict the blocks (personal and environmental) in their efforts to influence others and make attempts to overcome them.

A combination of high need for achievement and power may produce a self-centered autocrat. However, they do develop emotional bonds with people they work with. Their need for affiliations is low. This is so because if they establish emotional bonds, they are likely to end up working for people rather than for their organizations.

A sense of effectiveness' is yet another important psychological dimension that contributes to successful entrepreneurship. Entrepreneurs tend to present themselves as person striving towards goals that involve action. Being confident about their own abilities and resources, they are problem-solvers rather than problem-avoiders, as initiative takers rather as conformists. Their descriptions about themselves are likely to reflex a sense of confidence, a capacity for action-orientation. They tend to see themselves as effective person. Their responses to "Who am I" generally reflect these categories.

Entrepreneurs like challenge but do not take extreme risk. They experience challenges in undertaking tasks or making decisions that involve a moderate probability of success, and where they are sure that their efforts can influence their success. They do not like to undertake tasks which either very easy to accomplish or an on the other extreme which are impossible to achieve. The do not tend to like situations where the outcomes of a pursuit depends on chance and not on their efforts. They like to influence the outcome of their pursuit by making more efforts and then having a sense of accomplishment.

This orientation to moderate risk-taking amongst entrepreneurs become manifest in the efforts they make to ensure market for all their products before actually setting up their plant or going into production. Usually entrepreneurs spend a considerable amount of time planning their enterprises. They study the market situation, explore profitability in alternative lines of business, products, machinery, technology and process, finances, and compare before making their final decision. This indulgence in planning may be considered as an indicators of the calculated risk-taking behaviour on the part of the entrepreneur as it safeguards against subsequent difficulties, which can be anticipated.

Learning from experience and feedback may be considered as an equally important attribute of entrepreneurship. Successful entrepreneurs continuously modify their goals on the basis of the feedback they receive form their environment. In the "Toy Assembly Exercise" it is generally observed that if entrepreneurs undertaken to assemble a certain number of toys and realize that they could not accomplish it in the first, trial, they experience. They are basically inclined to test out their capabilities whenever an opportunity arises, and are open to feedback form such tests.

Experience with entrepreneurs indicates that their need for independence and their sense of determination are the two chief characteristics that drive them to start their won business and prefer not to be controlled by others. Very often they quote this as the reason for their desire to leaves service under some external agency and to set up their own business to fulfill their need for independence. As a consequence of their high need for independence, they may have had adjustment problems in their jobs in conformist organizations. Analysis of the family backgrounds of the entrepreneurs indicates that in their family environment as children, their individuality was reinforced and they enjoined ample freedom.

One to be positive view for his success. Aptitude for new enterprise and positive attitude give positive result. Entrepreneurs are inclined to approach their tasks with a hope of success. They attempt any task in the hope that they will succeed rather then with a fear of failure. Such a hope of success enhances their confidence.

Entrepreneurs also tend to believe in their own capacity to influence the environment. They would prefer to change it rather than leave everything to luck and God or to forces beyond their control. They believe that they can shape their own destiny, even though they might have faith in God and other forces and powers. Whenever they achieve something or get rewarded they tend to attribute it to their won efforts rather than to luck or to grace form someone else. In psychological terms, this is referred to as internal locus of control. Entrepreneurs have an internal locus of control. This helps them in putting in their efforts with vigour. They would like to experience that they can make things happen rather then see things happening by themselves. This also helps them to accept their mistakes and improve on them. When they fail, they tend to own their mistakes as they show inclinations to won their success.

A person with past orientations generally lives his past. Overwhelmed by past achievements and failures, he is unable neither to live present effective nor to project himself into the future.

In such a situation certain generalizations could, however, be made. One the traits described earlier can be developed in people through, psychological education. Two, presence of these traits increases probability of an entrepreneur emerging out successful which had a large number of achievement themes reflected in their literature, had economic development a few years later. This indicated hat if children get exposed to achievement themes and hear stories about characters striving for excellence; their thought processes get influenced. As a result, these children are likely to become entrepreneurs in their adult careers and thus contribute to economic development. An implication of this finding for education is to encourage achievement oriented literature in school curricula.

Person with high achievement came from where fathers were performing entrepreneurial roles in their occupations, regardless of whether the surrounding community was highly modern or traditional. There are several studies, which suggest that a good proportion of entrepreneurs emerge form families with entrepreneurial background business tradition.

What in a home of businessman that contributes towards building entrepreneurs? This has not been conclusively established by research in a business family, the parents tend to socialize their children towards self-reliance and independent, which might move an individual towards entrepreneurships. It is now known that much early socialization in the family plays a great role in drawing a person towards entrepreneurship.

A few researches have been conducted in India on agricultural entrepreneurs. These studies indicate that farmers with medium size of land holdings show greater motivation to achieve, and readiness to change, compared to both small and bigger landholders, and they younger ones are better disposed to achieve and change ones are better disposed to achieve and change, irrespective of the size of their land holding, this study also indicated that training and achievement motivation designed to help middle class and young farmers may offer better results. Other studies in this area as well as tend to underline the association between need achievement and entrepreneurial activity in farming. Several of the characteristics discussed in this chapter are measurable through psychological tests and exercises. Several such tests, exercises and scoring systems have been described elsewhere in this book. The assumption that people can be helped to develop characteristics that are entrepreneurial forms the basis of motivation development training. Sometimes people possess some of these characteristics without being away of them. Entrepreneurial motivation laboratories attempt to provide opportunities to develop these traits further. Experience has shown that such laboratories contributes in a major way to the effectiveness of a person. The rapidly increasing demand for these laboratories in most of the developing countries may be considered indicative of their usefulness.

REFERENCES

Extracted from the article of T. Venkateshwar Rao and Prayag Mehta, Entrepreneurship: The Concept and Social Context, (Ed) T.V. Rao and Udai Pareek, Developing Entrepreneurship: A Hand Book, 1978.

Mishra S.P Training from in Entrepreneurship NIESBUD, New Delhi Ojha, SN Profile of Entrepreneurs. (2000).

Chapter 15

ENTREPRENEURIAL PLAN FOR AQUARIUM BUSINESS

First of all the question arises, why plan is important.

❑ A plan of action.

❑ A blueprint for the future.

❑ Reduces uncertainty and "Seats of Pants" decisions.

❑ Lets you make mistakes on paper.

❑ Helps you identify your customers.

❑ Shows how much money you will need why.

❑ Highlights what skills and resources needed and why.

❑ Helps you communicate your ideas.

❑ Reduces the risk of failure.

❑ Gives you confidence the business will succeed.

❑ Tests your commitment and vision.

THE SIX PHASES OF BUSINESS PLANNING

Phase 1

History and position to Data

Introduction to the business

Description of the firm

Description of products & services

Phase 2

Collection of Data

Customers, market size & trends

Competitors

Market research project

Phase 3

Business Strategy

Pricing

Advertising & Promotion

Location

Selling

Manufacturing

Materials & Supply

Key personnel

Other elements

Phase 4

Forecasting Results

Sales forecast

Profit & Loss account

Cash flow

Balance sheet

Break even

Financing requirements

Phase 5

Business Controls

Book keeping Sales & Marketing other business controls

Phase 6

Writing up your Business Plan

ENVIRONMENTAL SCANNING SENSING BUSINESS OPPORTUNITIES

1. The need.
2. Entrepreneur's perspective.
3. What is an opportunities available.
4. Process-Generating Ideas–Short-listing–Final Selection.
5. Steps in idea generation.
6. Steps in idea generation.
7. Factors for short listing.
8. Final selection-Govt. Policy–SWOT Analysis.
9. Types of products
10. Types of markets/customers
11. Interplay of market Survey and product Selection.
12. Financial criteria
13. Product Evaluations. (Stability, Growth, Marketability, Enterprise related factors. Production capability).
14. Other criteria.

STEPS IN SETTING UP A SMALL INDUSTRIAL ENTERPRISE

The various steps involved in setting up a business enterprise will be most complex when it relates to an industrial unit. The step in setting up a small industrial enterprise shall be:

1. Deciding to go into business

This is the most crucial decision a youth has to take, shunning wage-employment and opting for self-employment/entrepreneurship.

2. Analysing Strengths/wcaknesses

Having decided to become an entrepreneur the young person has to analyze his/her strengths/weaknesses. This will enable him/her to know what type and size of business would be most suitable. This way vary from person-to-person.

3. Product Selection

The next step is to decide what business to venture into, the product or range of products that shall be taken up for manufacture and in what quantity. The level of activity will help in dividing size of business and form of ownership. One could generate a number of project ideas through environment scanning, short list a few items, closely examine each one of these and zero on to a final products.

4. Market Survey

It is easy to manufacture an item but difficult to sell. So it is prudent to survey the market before embarking up on production and satisfy the product chosen is in demand changes in product design required, determining demand-supply gap, extent of competition your potential share of the market, pricing and distribution policy etc.

5. Form of Ownership

A firm can be constituted as *proprietorship, partnership, limited company,* (public or private) *co-operative society* etc. This will depend upon the type purpose and *size* of your business. One may also decide on the form of ownership based on *resources on hand* or from the *point of saving on taxes.*

6. Location

The next step will be to decide on the place where the unit is to be located. Will it be hired or owned? The size of plot and covered and the last one identified. This will be useful in determining the machinery and equipment to be installed.

8. Machinery and Equipment

Having chosen the technology, the machinery and equipment required for manufacturing the chosen products have to be decided, suppliers identified and then costs estimated. One may have to plan well in advance for machinery and equipment especially if it has to be procured from outside the town, state or country that is, have to be imported.

9. Project Report

The *economic* viability and *technical feasibility* of the product selected has to be established through a project report. A project report that may now be prepared will be helpful in formulating the *financial, production,* marketing and management plans. It will also be useful in obtaining finance, shed, power, registration, raw material quotas etc.

10. Finance

Money is no problem for starting a small scale industry. But an entrepreneur has to take certain steps and follow specific procedures to obtain it. A number of financial agencies will give loans on concessional terms. Under TRYSEM and Schemes entrepreneurs are also eligible for subsidies which obviate the seed margin money.

11. Power Connection

The site chosen should either have adequate power connection or this should be arranged now.

12. Installation of Machinery

Having arranged finance, work shed, power etc. the next step is to procure the machinery and begins its installation.

13. To be determined, sources of getting desired labour identified and labours as start recruited. Possibly, the labour has to be trained either at the entrepreneur's premises or in a training establishment.

THERE'S ALWAYS A BETTER WAY : MICRO SCREENING

Setting

The output of Macro Screening consists in having a group or individual participants identify 20 potentially viable project ideas that the group or individual participant thinks feasible of undertaking in a specific locality dictated by the facilitators, by the location of the training, or by the origin of the participants from the hundreds of ideas that have been generated as a result of the *Brainstorming session.* The 20 identified potentially viable project ideas are put in a flip chart. A brief presentation by the group representatives of individual participants may be called for by the facilitator.

Note

No individual participant should be forced to present his project ideas as he/she may reason for their confidentiality. This is specially true, where the individual participants stick to their original projects as mentioned by them in their application forms or during the interviews. The participants should however, be open to the facilitators to receive guidance and advice in the development of their business plans.

During the appreciation workshop or *training of trainers (TOT) program,* groups can be called by the facilitator for class presentation.

After the presentation by the participants, the facilitator explains that the 20 potentially viable project ideas will under go a *finer screening* test, that is the *Micro Screening.* He then proceeds to explain the concept. He uses a readymade. *Micro Screen chart* (see illustrative chart) containing various columns with the following headings:

PJ	MKT	RM	TEC	SKL	GOP	SFT	EI	RE	PFT	C/B	TTL	CF
1												
2												
3												
4												
5												
6												
7												
8												
9												
10												

14. Raw Materials

The labour will require raw material to work upon the installed machines raw materials required may be available indigenously or may have to be *imported Government agencies can assist it the raw materials* required are scare or imported.

15. Production

The unit established should have an organisational set up. That is, the structure of the man power proposed to be *employed must* be determined. This ensure *smooth* and effective running of the unit. There should not be any wastage of man power, material or machine capacity installed. If the items produced are exported then the product and its packaging must be attractive.

Production of the proposed items should be taken up in two stages :

(*a*) Trial production

(*b*) Commercial production.

Trial production will help *teaching problem confronted in* production and *test marketing of the products*. This will reduce chance of losses in the eventuality of mistakes in project conception. Only after *successfully launching* the product at test market stage should commercial production be commenced.

16. Marketing

Having manufactured the product, the stage comes to sell it. This is called marketing, Various aspects like how to reach the customer, distribution channels, commission structure, pricing, advertising/publicity etc. would have already been decided upon at market survey/ project formulation stage.

Like production, marketing should initially be attempted continuously, that is, in two stages:

(*a*)   Test Marketing.

(*b*)   Commercial Marketing.

Test Marketing will save the enterprise from going into disrepute just in case product launched is not *well accepted* in the market, it will also assist in carrying out *modifications/ characteristics/features* of the product.

Having successfully test marketed the products commercial marketing can be undertaken.

17. Quality

After or, at times, before marketing the product, quality certification like BIS, 'Q' Mark, Agmark, 'Del-in' etc, depending upon the products should be obtained. If there are no quality standards specified for the products the entrepreneur should evolve his/her own *quality control parameters*. Quality, after all, ensures long-term success.

Column

1.	Project	(PROJ)
2.	Availability of Market.	(MKT)
3.	Availability of Raw Materials.	(RM)
4.	Availability of Technology.	(TEC)
5.	Availability of Skills.	(SKL)
6.	Government Priority.	(GOP)
7.	Strategic Fit	(SFT)
8.	Ease of Implementation	(EI)
9.	Risk Exposure	(RE)
10.	Profitability	(C/B)
11.	Cost/Benefit	(PCB)
12.	Total	(TTL)
13.	Critical Success Factor	(CF)

The 10 factors (starting with column 2 to column 11) will be rated according to the participant's or group's perception, availability of information, or experience in the following manner:

1. poor
2. fair
3. satisfactory
4. very satisfactory
5. excellent

The potential total score a project idea can get is 50 points, that is 5 for excellent multiplied by 10 factors.

The facilitator must explain the meaning and significance of the 10 factors using a specific project (that participants) as an example. The following is a guide :

1. *Availability of Market*

If there is no *assurance of an adequate market,* there is no sense going to business. The *market must be large enough* to enable the entrepreneur to capture a market share and in the process, make an *attractive profit* from the transaction after deducting his costs from sales. Indication of availability of market includes, among others:

(a) existing *demand* is not at present adequately served by existing suppliers;

(b) existing demand is presently served by imports;

(c) existing demand is *presently not served at all;*

(*d*)   the *project's product* has *significant uniqueness* or unique selling features–USF (in the case of a product) or unique selling proposition–USF (in the case of a service) such as more desirable features, better quality, more durable, better taste, superior after sales service, free delivery, etc.;

(*e*)   *supply of the product/service is not reliable;*

(*f*)   demand for the product/service is expected to increase significantly or substantially in the *future;*

(*g*)   supply of the product is presently served through *smuggling.*

2. *Availability of Raw Materials*

Availability of raw materials is indicated by the following considerations :

(*a*)   raw materials are available in *adequate quantity locally;*

(*b*)   there is *reliability of supply* whether local or imported source;

(*c*)   seasonality, perishability, quality and variability of raw materials have been considered and found to be satisfactory;

(*d*)   *price* of raw materials is *reasonable;*

(*e*)   increases in the price of raw materials in the future is perceived to be *reasonable* and *predictable.*

3. *Availability of Technology*

Availability of technology can be evaluated in terms of the following indicators:

(*a*)   the technology or technologies to be used have been *proven;*

(*b*)   *reasonably priced technologies* are available to produce the product;

(*c*)   technology is *appropriate for the level of production,* level of *investment* and *desired product quality;*

(*d*)   the project will not suffer from *technology obsolescence* which will render the project not viable.

4. *Availability of Skills*

Availability of skills can be gauged by the following factors:

(*a*)   *different skills* (conceptual, managerial, technical and manual) needed by the project are available;

(*b*)   *supply of skills is relatively steady and stable* so as not to jeopardize the project in case of sudden or unforeseen labour changes usual turnover or unexpected problems;

(*c*)   *cost of labour* is projected to be fairly steady and predictable.

5. *Government Priority*

Government priority is indicated by the following considerations :

(*a*)   the project is listed under the *governments list of priorities* for promotion or investment;

(*b*)   the project receives *government incentives whether* fiscal (*e.g.* tax) exemption or reduction, tariff protection, import privileges, etc) monetary (priority lending status, reduced interest rate), or other *support assistance* (*e.g.* marketing, technical or consultancy services);

(*c*)   the project falls under the government's priorities of *import substitution, export promotion, employment generation, rural industrialization programs, technology development/transfer, etc.*

6. *Strategic Fit*

The strategic fit of the project under consideration is indicated by the following criteria :

(*a*)   proposed project fits well within the *competence and expertise* of the entrepreneur or key staff;

(*b*)   proposed project fits well within the *existing product line*, technology, marketing set up, production system, facilities and resources of the entrepreneur or the firm;

(*c*)   project complements and enhances the existing set up, viability or *growth of the firm through* a positive synergy.

7. *Ease of Implementation*

Ease of implementation can be measured by the following criteria :

(*a*)   project can *easily be implemented* because the inputs are readily available;

(*b*)   project can be implemented within a *short gestation period* or *reasonable preparatory period* (*e.g.* 3 *months to one year*);

(*c*)   project can *start operation within one year form the completion* of the training;

(*d*)   any *unforseen difficulties can be controlled* by the *entrepreneur* or management.

8. *Risk Exposure*

The project is rated excellent if it is considered less risky, or risks are very minimal while projected profits are more or less assured. Degree of risk can also be assessed in terms of the following factors :

(*a*)   the product or service can readily be *copied or imitated* if the project is found very profitable by others;

(*b*)   competitiors who have more *resources and expertise may effectively fight back if threatened by the project;*

(*c*)   changes in *customer's and consumers'* lifestyle, buying habits, consumption and spending pattern, etc., may take place anytime before the project can service the market;

(*d*)   the project may suffer from unforeseen factors such as weather condition, availability of raw materials, technology obsolescence, changes in government policies, priorities or programs;

(*e*)   dependency of the *project on imported inputs* whether raw materials, technology, skills or other resources.

9. *Profitability*

At this point in the training, the participants have not yet undertaken business plan preparation. In this case, profitability is a matter of perception or prior knowledge of the participants which may be based on their readings, experience, or observations/feedback from others (competitors, suppliers, consultants, among others). This criterion can be rated in terms of whether the profitability (return on investment) is :

(*a*)   excellent

(*b*)   very satisfactory

(*c*)   satisfactory

(*d*)   fair

(*e*)   poor

10. *Cost/Benefit*

This tenth factor is practically the summation of all the other nine criteria and gives as overall impression regarding the desirability and feasibility of undertaking the project. The various considerations to be taken into account in rating this factor include :

(*a*) whether the benefits of the project in terms of *profitability*, risk, *investment requirement, availability* of inputs, etc, are worth all the efforts in conceptualizing, organizing, and implementing the project.

(*b*) the project provides *sufficient* (tangible and visible) benefits to the community-either through employment generation, backward or forward linkages with other industries or other economic activities availability of needed products and services, etc.

(*c*) the project is economically viable through its own merits and not through artificial government interventions.

*The **Critical Success Factor** (CSF)* concept can then be explained. CSF means a creation factor *particular to the identified project which is very important to the success of that specific project.* If that certain factor is missing, is inadequate, or is not properly taken into account, it can suggests that the project is not *feasible* or is *likely to collapse.*

A project's **CSF** can be one of *the 10 factors above.* However, it must be *further refined or qualified.* For example, not just raw materials, but seasonality of raw materials, **perishability** of raw materials, lack of **standardization** of raw materials, **unpredictability of availability of supply** (perhaps due to import restrictions, infrastructure problem, weather condition, etc.).

The CSF concept is also a way of validating the participant's explanation and rating with regard to the 10 micro screening factors.

After the explanation of the whole Micro-screening technique (the meaning of the 10 factors, the rating system, and the CSF), the facilitator can give the participants about 30 minutes to one hour to go through the exercise-asking the participants to micro screen the 20 identified project ideas. The facilitators should be around to provide further explanations to the 10 criteria or guidance to the participants.

To allow more time for discussion or evaluation, the facilitator can prepare in advance the Micro-screen chart according to the number of participants or groups.

After the exercise, the facilitator can ask for class presentation. In a SPIW or full course setting, volunteer participants can explain the three highest rated projects. In an appreciation workshop or training of trainers, the Micro-screen is a group process, hence each group can be called by the facilitator to take turns in presenting their three highest rated project ideas. In any case, each presentation by a volunteer participant or group should not take more than 10 minutes. The class should be encouraged to ask questions to help crystallize or clarity the participants' ratings. It has been the experience that the other participants always provide constructive feedback particularly providing information heretofore unknown or unclear by the presenting participant or group.

The Micro-screen leads to the SWOT (Strengths, Weaknesses, Opportunities and Threats) session.

Before making/submitting a project one should see the following aspects also :

DEMAND

The quantity of a product or service that buyers will purchase at different prices in a given market of a given time.

Demand Schedule

The tabular expression of quantity demanded at different price level.

Demand Curve

The graphical presentation of price and quantity demanded pairs on a two dimensional space.

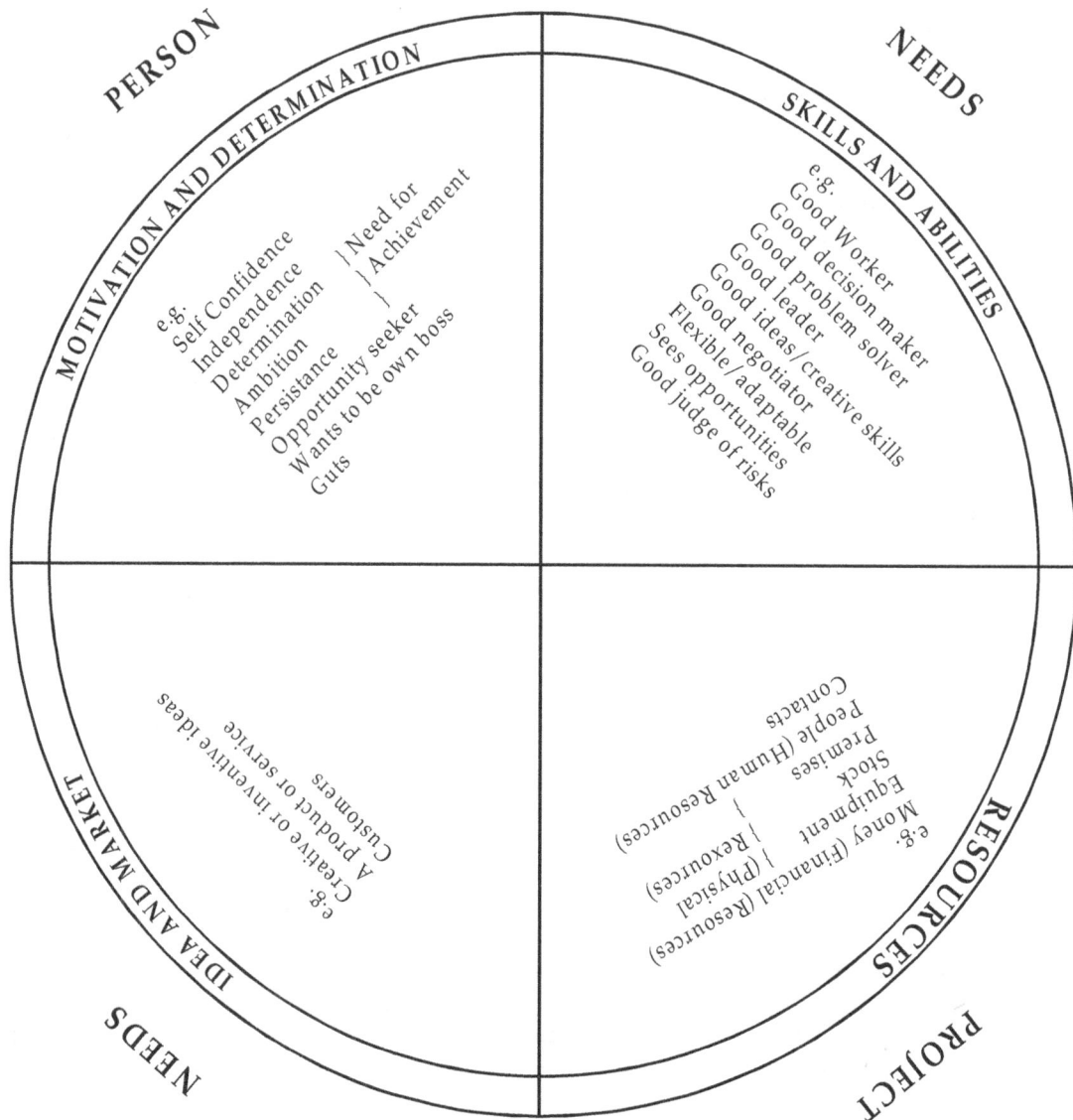

Diagram — circular chart with four quadrants:

Outer labels: **PERSON**, **NEEDS**, **SKILLS AND ABILITIES**, **MOTIVATION AND DETERMINATION**, **IDEA AND MARKET**, **NEEDS**, **RESOURCES**, **PROJECT**

Top-left quadrant (MOTIVATION AND DETERMINATION):
e.g.
Self Confidence
Independence } Need for
Determination } Achievement
Ambition
Persistance
Opportunity seeker
Wants to be own boss
Guts

Top-right quadrant (SKILLS AND ABILITIES):
e.g.
Good Worker
Good decision maker
Good problem solver
Good leader
Good ideas/creative skills
Good negotiator
Flexible/adaptable
Sees opportunities
Good judge of risks

Bottom-right quadrant (RESOURCES):
e.g.
Money (Financial Resources)
Equipment } Physical
Stock } Resources
Premises
People (Human Resources)
Contacts

Bottom-left quadrant (IDEA AND MARKET):
e.g.
Creative or inventive ideas
A product or service
Customers

Demand Function

The mathematical expression of the relationship between the quantity demanded (of a commodity or group of commodities) and factors affecting the quantity demanded.

$$D = f\,(P_t,\ P_{st},\ Y_t,\ T_t,\ S_t,\ E_t,\ D_{t-1}),\ \text{where}$$

D = Demand-Quantity of fish demanded in period t.

P_t = Price of fish in period t.

P_{st} = Price of substitute like meat in period t.

T_t = Taste and preferences in period t.

Y_t = Average income level in period t.

S_t = Size of the population in period t.

E_t = Average level of education in period t.

D_{t-1} = Quantity of fish demanded in period t-1.

SUPPLY

The quantity of a product or service that are offered for sale at different prices in a given market at a given time.

Supply Schedule

The tabular expression of quantity supplied at different price level.

Supply Curve

The graphical presentation of price and quantity supplied pairs on a two dimensional space.

Supply Function

The mathematical expression of the relationship between the quantity supplied (of a commodity or group of commodities) and factors affecting the quantity supplied.

$S = f (P_t, P_{st}, Y_t, T_t, E_t, S_{t-1})$, where

S = Quantity supplied in period t.

P_t = Price of substitute like of meat in period t.

P_{st} = Price of fish in period t.

Y_t = Average income level in period t.

T_t = Taste and preferences in period t.

S_t = Size of the population in period t.

E_t = Average level of education is period t.

S_{t-1} = Quantity of fish supplied in period t-1.

Equilibrium Price

The price at which the quantity demanded equals the quantity supplied.

$E_p = Q_d = Q_s$

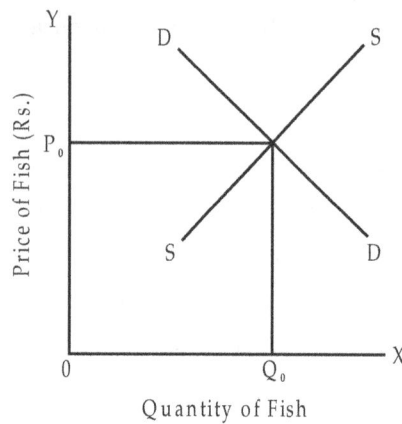

Example

For the given set of price data and quantities estimate the equilibrium point.

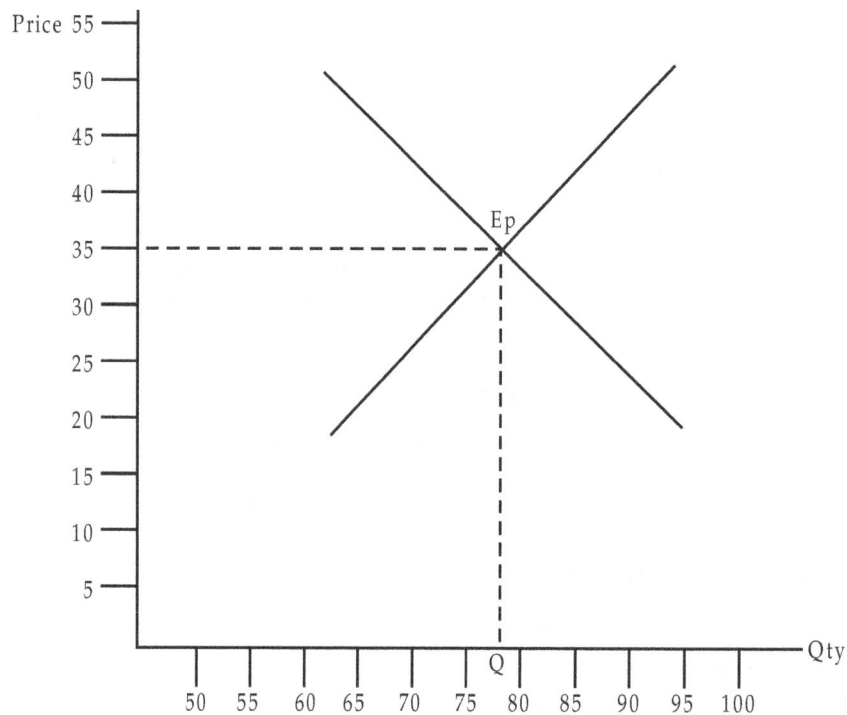

Factor-Product Relationship (Input-Output Relationship)

The nature of relationship between input and output can be classified as:

❑ Law of increasing returns (Increasing marginal productivity)

❑ Law of decreasing returns (Decreasing marginal productivity).

Among the three laws, law of decreasing returns operates and founds applicability in the case of aquaculture.

❑ Law of decreasing returns

❑ Law of diminishing marginal returns

❑ Law of constant returns (constant marginol productivity).

An increase in capital and labour applied in the cultivation of land in general causes, less than proportionate increase in the amount of product raised, unless it helps to coincide with an improvement in the art of agriculture (Marshall).

1. Total Product (TP)

The sum total of the product produced by the factors of production. It is usually denoted by Y.

2. Average Product (AP)

The average output is defined as the total product divided by the total input. AP=Y/X

3. Marginal Product (MP)

The additional output obtained by one more unit addition of input.

4. Irrational Production

There exists a possibility for the rearrangement of input levels either to get a greater product from the same level of resources or to get the same product with smaller outlay or resource levels. It can be also be due to employment of inefficient technique that does not permit the greater output as above. The first and second stages of production denotes the irrational stages of production.

5. Rational Production

The production level where the product and input cost puts the economic issue into proper perspective. Usually denoted in the second stage of production.

6. Elasticity of Production

Indicates responsiveness of changes in output over change in input.

$$E_p = \frac{\% \text{ change in output}}{\% \text{ change in input}} = \frac{\dfrac{\Delta Y}{Y} \times 100}{\dfrac{\Delta X}{X} \times 100} = \frac{\Delta Y}{\Delta X} \Big/ \frac{Y}{X} = \frac{MP}{AP}$$

7. Profit Maximisation

Marginal Product = Factor – Product Price Ratio

$$\frac{\Delta Y}{\Delta X} = \frac{P_x}{P_y}$$

8. Value of Marginal Product (VMP)

VMP = Marginal Product × Price of Output

   = MP × P$_y$

9. Over use, Under use, Optimum level of input

$$\text{VMP} > \text{P}_x \quad \text{Underuse}$$
$$\text{VMP} < \text{P}_x \quad \text{Overuse}$$
$$\text{VMP} = \text{P}_x \quad \text{Optimum}$$

10. Elastic, Inelastic and Unitary Elasticity of Production

Elastic	=	More than one
Inelastic	=	Less than one
Perfectly elastic	=	Infinity
Perfectly inelastic	=	Negative
Unitary elastic	=	One

11. Point of Inflexion

Point on the total product curve where the rate of increase in output start to fall of and the production function becomes concave downwards. When MPP is maximum, the corresponding on Total Product Curve is called point of inflexion.

Example

Determine the following with the given data :

(a) Three stages of production.

(b) Rational and irrational stages.

(c) Elasticities.

(d) Average Product, Marginal Products.

Input	Output	Change in input	Change in output	Average Product	Marginal Product	Elasticities	Remarks
0							
1							Unitary elastic
2							Elastic
3							Elastic
4							Elastic
5							Elastic
6							Inelastic
7							Inelastic
8							Inelastic
9							Inelastic
10							Perfectly inelastic
11							Perfectly inelastic

Graphical representation depicting the three stages of production

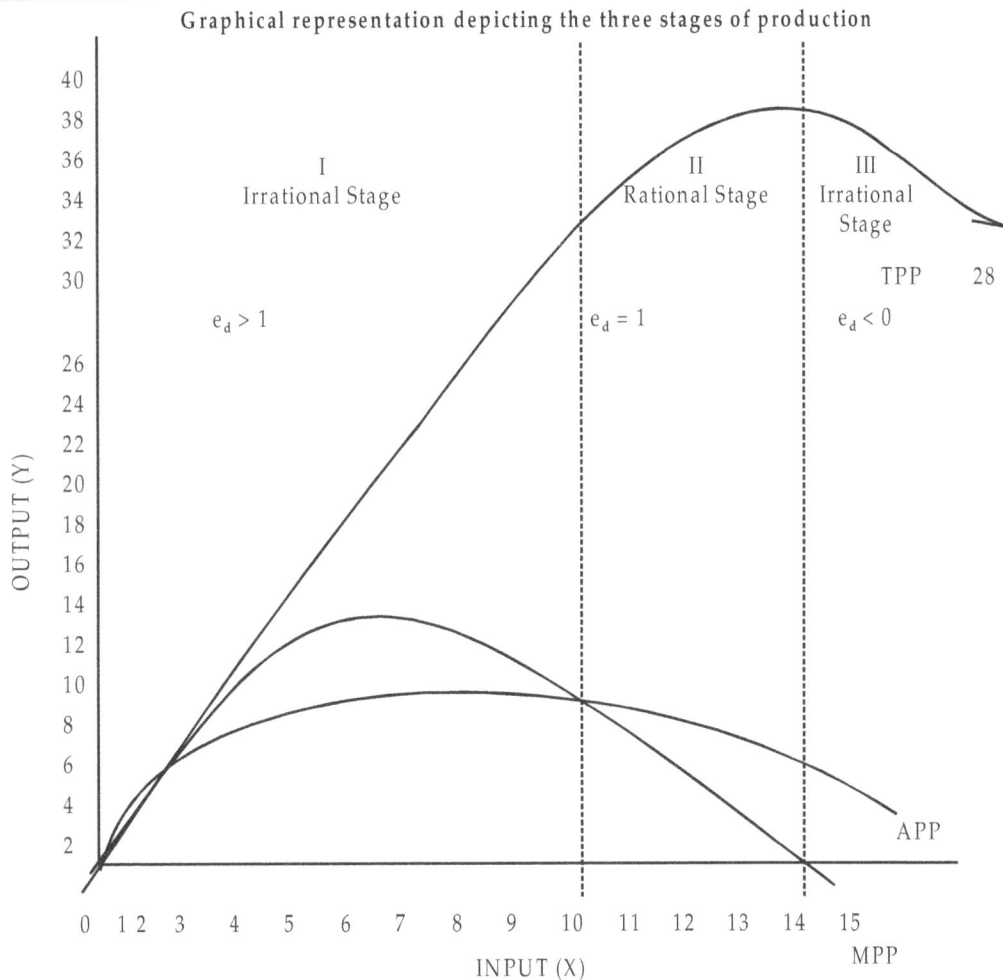

INPUT-INPUT, RELATIONSHIP

Law of Substitution

When more than one means of producing a given result is known and available, the least costly will be selected.

Objectives

☐ Minimisation of cost to a given level of output of a product.

☐ Optimisation of output to the fixed producing unit through alternative resources use combination.

Isoquants/Isoproducts/Product Indifference Curves

All possible combinations of two inputs, physically capable of producing the same amount of output. For example, they are convex to origin and downward sloping. The slope indicates rate of substitution between two inputs.

Marginal Rate of Technical Substitution (MRTS)

Rate at which two resources can be substituted, *i.e.,* how much the use of one resource can be reduced in order to add additional unit of the other factor to maintain the same level of output.

Isoquants at different output level

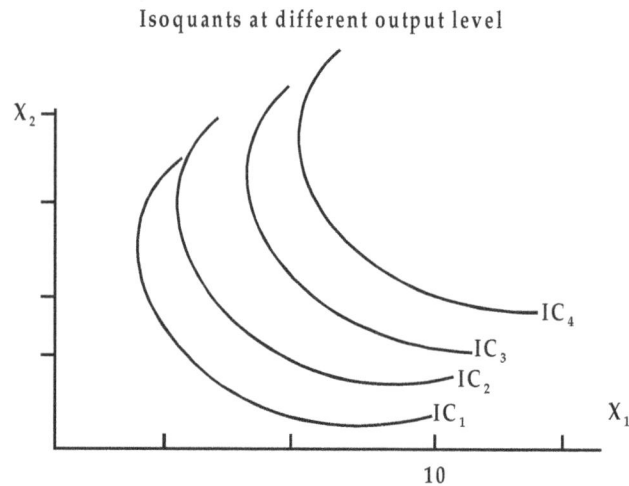

Price line at different price outlays

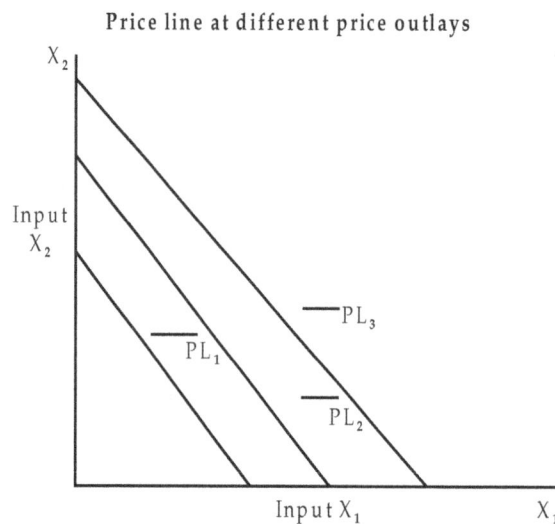

$$MRTS_{X_1X_2} = \frac{\Delta X_2}{\Delta X_1} = \frac{\text{No. of units of replaced resource}}{\text{No. of units of added resource}}$$

Elasticity of Substitution (E$_s$)

$$E_s = \frac{\text{Percentage change in one factor}}{\text{Percentage change in other factor}}$$

$$= \frac{\dfrac{\Delta X_1}{X_1} \times 100}{\dfrac{\Delta X_2}{X_2} \times 100} = \frac{\dfrac{\Delta X_1}{X_1}}{\dfrac{\Delta X_2}{X_2}} = E_s$$

$$= \frac{\Delta X_1}{\Delta X_2} \times \frac{\Delta X_2}{\Delta X_1} = E_s$$

Isocostline / Equal Cost line / Price line / Outlay line

All possible combinations of two inputs that can be purchased at a given cost.

$$\text{Slope} = \frac{P_{X_1}}{P_{X_2}}, \text{ where}$$

P_{x_1} = price of input X_1

P_{x_2} = Price of input X_2

Least Cost Combination

If two or more factors are employed in production of a single product, cost is at a minimum when the ratio of factor price is inversely equal to the marginal rate of substitution of factors.

Algebraic Expression of Cost Minimisation

1. *Computer Marginal Rate of Substitution Ratio*

$$MRS_{X_1 X_2} = \frac{\Delta X_2}{\Delta X_1} = \frac{\text{No. of units of replaced resource}}{\text{No. of units of added resource}}$$

2. *Compute Price Ratios*

$$\frac{P_{X_1}}{P_{X_2}} = \frac{\text{Cost per unit of added ratio}}{\text{Cost per unit of replaced ratio}}$$

3. *Least Cost*

$$MRS_{X_1 X_2} = \text{Price Ratio}$$

$$\frac{\Delta X_2}{\Delta X_1} = \frac{P_{X_1}}{P_{X_2}}$$

$$P_{X_1} \times \Delta X_1 = P_{x_2} \times \Delta X_2$$

Cost of adding X_1 is equal to reduction in cost from X_2.

$$\frac{P_{X_1}}{P_{X_2}} < \frac{\Delta X_2}{\Delta X_1} = \text{Cost can be lessened by using more of } X_1 \text{ and less of } X_2$$

$$\frac{P_{X_2}}{P_{X_1}} < \frac{\Delta X_1}{\Delta X_2} = \text{Cost can be lessened by using more of } X_2 \text{ and less of } X_1$$

Isoclines

A line or curve connecting the least cost combination of inputs for all output levels. Isoclines connect equal slope or equal MRS.

Ridge Lines

Lines, which separate complementary from substitution.

Expansion Path

Isocline for a particular production period at particular price level and input required.

Example

For the given data regarding the use of organic manure (X_1) and fertilizers (X_2) to produce a fish of 500 kg, compute the following:

(*a*) Marginal rate of substitution

(*b*) Elasticity of substitution

(*c*) Price ratio

(*d*) Least cost combination point

(*e*) Isoquant

Input	X_1	ΔX_1	X_2	ΔX_2	y	MRTS $= \dfrac{\Delta X_2}{\Delta X_1}$	Elasticity E_s	Price of X_1 @4	Price of X_2@1	Total Outlay of $X_1 \& X_2$	Remarks
A	1		12		40	—	$\dfrac{\Delta X_2}{\Delta X_1} \times \dfrac{X_1}{X_2}$	4	12	16	
B	2	1	8	4	40	4	2	8	16	24	Least cost combination
C	3	1	5	3	40	3	1.8	12	20	32	
D	4	1	3	2	40	2	2.66	16	12	28	
E	5	1	2	1	40	1	2.50	20	8	28	

OUTPUT–OUTPUT RELATIONSHIP

Basic Concepts

❏ Law of equal-marginal returns.

❏ Profits are maximised by using resources in such a way that the marginal returns from that resources are equal in all cases.

Objectives

❏ Profit maximisation with a given resource allocation when two or more products are being produced.

❏ Best combination of product for a given outlay of resources.

Production Possibility Curve/Transformations Curves/Iso-resource Curves/Iso-factor Curves

All possible combination of two products, when an equal quantity of resources is available for the two products.

They indicate the opportunities or possibilities in the production of two enterprises when resources are constant.

The production possibility curve is concave to the origin. The slope indicates rate of substitution between two outputs.

Marginal Rate of Product Substitution (MRPS) / Marginal Rate of Product Transformation (MRPT)

Rate at which two outputs can be substituted, i.e., the rate of change in quantity of one product as a result of a unit increase in the other product given that the amount of the inputs used remains constant.

$$MRPS_{Y_1Y_2} \frac{\Delta Y_2}{\Delta Y_1} = \frac{\text{Number of units of reduced output}}{\text{Number of units of added output}}$$

Elasticity or Product Substitution

$$E_{PS} = \frac{\text{Percentage decrease in output of one product}}{\text{Percentage increase in output of second product}}$$

$$= \frac{\Delta Y_2}{Y_2} \bigg/ \frac{\Delta Y_1}{Y_1} = \frac{\Delta Y_2}{\Delta Y_1} \times \frac{Y_1}{Y_2}$$

Isorevenue Line

❑ Indicates ratio of prices for the two competing products.

❑ Line that defines all the possible combinations of two commodities, which would yield an equal revenue or income.

$$\text{Slope} = \frac{P_{Y_2}}{P_{Y_1}}, \text{ where}$$

P_{Y_1} = Price output Y_1 for one unit

P_{Y_2} = Price output Y_2 for one unit

Maximun Profit Combination

The computation of maximum profit combination is by the following method.

(i) $MRPS_{Y_1Y_2} = \dfrac{\Delta Y_2}{\Delta Y_1}$

(ii) $\text{Isorevenue line} = \dfrac{Py_1}{Py_2}$

(iii)   Maximum Profit Combination

$$MRPS_{Y_1Y_2} = \text{Isorevenue line}$$

$$\frac{\Delta Y_2}{\Delta Y_1} = \frac{Py_1}{Py_2}$$

$$P_{Y_1} \times \Delta Y_1 = P_{Y_2} \times \Delta Y_2$$

Example

For the given data find the following :

1. Marginal rate of product substitution
2. Elasticity of product substitution
3. Maximum profit combination
4. Production possibility curve
5. Isorevenue line

Input X	Output Y_1	Product Y_2	Marginal Change in output ΔY_1	Marginal Change in output ΔY_2	MRPS $\Delta Y_2/\Delta Y_1$	MRPS $\Delta Y_1/\Delta Y_2$

Example

For the different levels of inputs and output, compute:

1. Total Cost (TC)
2. Average Cost (AC)
3. Average Fixed Cost (AFC)
4. Average Variable Cost (AVC)
5. Marginal Cost (MC)

Also represent lines graphically.

Output	FC	VC	TC	AC	AFC	AVC	MC
0							
1							
2							
3							
4							
5							
6							
7							
8							
9							
10							

Inferences

- ❑ Average Fixed Cost decreases with increasing output.
- ❑ Average Variable Cost decreases and after a particular level (MP = 0) the Average Variable Cost again increases.
- ❑ The condition of no loss or profit.
- ❑ Cost allocated to a product is equal to all revenues from its sales.
- ❑ Smaller quantity yields losses.
- ❑ Higher quantity yields profit.

Computation

- ❑ Algebraic Method.

  Computation of BEP in units.

$$\underset{\text{(Quantity)}}{\text{BEP}} = \frac{F}{P - V}, \text{ where}$$

  F = Fixed cost on Rupees per hectare of fish farm.

  P = Price per quintal of fish in Rupees.

  V = Variable cost per quintal of fish in Rupees.

$$\text{Break even point in monetary value} = \text{BEP} = \frac{F}{\left(1 - (V/P)\right)}$$

Margin of Safety

- ❑ Difference between total output and output at BEP.
- ❑ Difference between total revenue obtained from the enterprise and revenue at BEP.

$$\text{Margin of Safety} = \text{Total Output} - \text{Output at BEP}$$

$$= \text{Total revenue} - \text{Revenue at BEP}$$

$$\text{Percentage margin of safety (Qty)} = \frac{\text{BEP output}}{\text{Volume of output}} \times 100$$

$$\text{Percentage margin of safety (In money)} = \frac{\text{BEP in monetary value}}{\text{Total revenue}} \times 100$$

Example

Estimate the profits of the two aquaculture farm.

Sl.No.	Fixed Cost	Variable Cost	Total Cost	Price per Quintal (Rs.)	Volume of Output (Qt)	Total Revenue (Rs.)	Variable Cost Fund (Rs.)
1							
2							

Farm I

$$BEP(Qty) \longrightarrow \frac{F}{(P - V)}$$

$$BEP(Monetary\ Value) \longrightarrow \frac{F}{(1 - V/P)}$$

$$Margin\ of\ Safety \longrightarrow Total\ Output - Output\ of\ BEP$$

$$Margin\ of\ Safety = \frac{BEP\ output}{Volume\ of\ output} \times 100$$

$$Margin\ of\ Safety\ (in\ money) = \frac{BEP\ in\ Monetary\ value}{Total\ revenue} \times 100$$

Farm II

$$BEP(Qty) \longrightarrow \frac{F}{(P - V)}$$

$$BEP(Monetary\ Value) \longrightarrow \frac{F}{(1 - (V/P))}$$

$$Margin\ of\ Safety \longrightarrow Total\ Output - Output\ of\ BEP$$

$$Margin\ of\ Safety = \frac{BEP\ output}{Volume\ of\ output} \times 100$$

$$Margin\ of\ Safety\ (in\ money) = \frac{BEP\ in\ Monetary\ value}{Total\ revenue} \times 100$$

Inference

Positive Margin of Safety indicates shock absorbing capacity of the enterprise in the event of fluctuation in returns against anticipation owing to any unforeseen eventualities of risk and uncertainties.

Production function is a mathematical and functional relationship showing the transformation of inputs to outputs.

Basic Form

1. Linear production function $Y = a + bx$
2. Cobb-Douglas (Log linear) $Y = a \cdot x^b$
3. Quadratic $Y = a + bx - cx^{2}$

where Y – Dependent variable

x – Independent variable

The choice of the functional farm depends upon the basis of Scatter diagram (reveals the relationship between the input data and output data and indicates the overall trend).

(i) Linear Form

General form: $Y = a + bx$, where   Y – Output in unit

a – Intercept

b – Slope/Marginal product

x – Independent variable in unit

(ii) Cobb – Douglas

Log – log form

General form $Y = a \cdot x^b$

$\Rightarrow$ $\log Y = \log a + b \log x$, where   Y – Output in unit

a – Intercept

x – Independent variable in unit

b – Elasticities

General form $Y = a + b_1 x_1 + b_2 x_2 + \dots + b_n x_n$

$Y = a \cdot x_1^{b1} \, x_1^{b2} \, x_3^{b3} \, \dots\dots \, x_n^{bn}$

Computation Procedure

Convert the data into per hectare or per unit area.

(i) Zero order correlation matrices to find the determining variables and deleting the variables possessing multi-colinearity.

(ii) Run the regression using the data having input and output variables (MS EXCEL, SPSS).

(iii) Estimation and interpretation of the results.

(a) R² (Coefficient of Multiple Determination)

The percentage variation in the output as explained by the selected independent variables.

$R^2 = 0.58$, indicates that 58 per cent of the variation in Y is explained by the selected independent variables.

(b) Marginal Product/Slope

The change in output by an additional unit of input. Coefficient in the summary sheet of the result.

(c) Average Product

$$\text{The Average product is} = \frac{Y}{X}$$

$$\text{For 'n' inputs, AP} = \sum_{x=1}^{n} \frac{Y}{X_n}$$

(d) Elasticities $= \dfrac{\textbf{Marginal Product}}{\textbf{Average Product}}$

Based on which, elastic, inelastic, unitary elastic can be worked out.

(*e*) Value of Marginal Product (VMP)

VMP = Marginal Product of Output x Price of Output

$$= MPP_y \times P_y$$

(*f*) Usage of Inputs

Underuse	VMP > P_x (price of input)	
Overuse	VMP < P_x	
Optimum	VMP = P_x	

Significant Variables

Determining Yield

Based upon the 'F' value, 't' value. The significance are estimated for 1% and 5% level of significance based on the table value.

(*g*) Returns to Scale

The sum of the elasticity (e_d) values yields the returns to scale.

$e_d > 1$ Increasing returns to scale

$e_d < 1$ Decreasing returns to scale

$e_d = 1$ Constant returns to scale

Marketing

Marketing is the performance of business activities that direct the flow of goods and services from producer to consumer or user.

Marketing Functions

Any single activity performed in carrying a product from the point of its production to the ultimate consumer may be termed as a marketing function.

It may have any one or combination of three dimensions, *viz.*, time, space and form.

E.g., The marketing of fish may involve carrying, price determination, selling, buying, grading, processing, packing, storage, etc.

Marketing Channels

Marketing channels are routes consisting of intermediaries through which commodities move from producers to consumers.

E.g., Fish marketing channels

(*i*) Fisherman - Auctioneer - Retailer - Consumer

(*ii*) Fisherman - Auctioneer - Processor - Wholesaler - Retailer - Consumer

Price Spread

The difference between the price paid by consumer and the price received by the producer for an equivalent quantity of product is know as Price spread. Marketing system is efficient when price spread is minimum. The price spread includes :

(*i*) **Marketing Cost (MC) :** The costs or expenses incurred in moving the product or service from producers to consumers.

E.g. - Transportation, packing, processing, etc.

(*ii*) **Marketing Margin (MM) :** Profits or income earned by various market intermediaries involved in moving the produce from the production to the ultimate consumption.

E.g. - Commission, retailer's profit, etc.

So, Price Speed = Consumer's Price (CP) – Producer's Price (PP)

= Marketing Cost + Marketing Margin.

Example

Case I

A fisherman, Mr. Rajhans comes to Versova Fish Landing Centre, with 1.00 kg of Sciaenid fish. The transportation charges to bring the fishes to the landing center is @ Re. 0.50/kg, He takes the fishes to an auctioneer, Mr. Ramlal and the fishes are auctioned and one wholesaler Mr. Chandi purchases the lot @ Rs. 40/kg. Mr. Ramlal takes auction rate @ Rs. 0.25/kg form Mr. Rajhans. Mr. Chandi brings the fishes to Dadar market with transportation cost @ Re. 0.75/kg and sells the lot to a retailer, Mr. Amit @ Rs. 45/kg. Mr Amit sells the fishes to consumer @ Rs. 50/kg in the same market. It is assumed that there is no loss in transit and no significant time lag.

Case II

A fisherman, Mr. Kasim comes to Versova Fish Landing Center, with 100 kg of Sciaenid fish. The transportation charges to bring the fishes to Versova landing center is @ Re. 0.50/kg. He takes the fishes to an auctioneer-cum-retailer, Mr. Rama and sells the lot @ Rs.40/kg. Mr. Rama then sells the fishes to consumers @ Rs.45/kg at Andheri market and he provides the transportation charges from Versova to Andheri market @ Re. 0.60/kg and icing charges @Re. 0.40/kg.

Work out MC, MM, Price Spread, and Producer's share in consumer's rupee and interpret for both the cases.

Solutions

Case I

(*a*) Transportation cost paid by Mr. Rajhans, fisherman

= 100 × 0.50 = Rs. 50.

(*b*) Transportation cost paid by Mr. Chandi wholesaler

= 100 × 0.75 = Rs. 75.

∴ Total Marketing Cost (MC) = A + B = 50 + 75 = Rs. 125

(*c*) Commission taken by Mr. Lalu auctioneer from Mr. Rajhans

= 100 × 0.25 = Rs. 25.

(*d*) Profit earned by Mr. Chandi

= 100 × {45 – (40 + 0.75)} = 100 (45 – 40.75)

= Rs. 425.

(e) Profit earned by Mr. Amit retailer

$$= 100 (50 - 45) = 100 \times 5 = Rs. 500.$$

∴ Total Marketing Margin (MM) = C + D + E = 25 + 425 + 500

$$= Rs. 950.$$

∴   Price Spread for 100 kg Sciaenid here

$$= MC + MM$$

$$= 125 + 950 = Rs. 1,075.$$

∴   Total price received by Mr. Rajhans, fisherman

$$= (100 \times 40) - (50 + 25) = 4000 - 75$$

$$= Rs. 3,925$$

∴   Total price paid by the consumers = 100×50 = Rs. 5,000

∴   Producer's share in consumer's rupee

$$= \frac{3925}{5000} \times 100 = 78.5 \text{ per cent}$$

Identification of Critical Activity

An activity can be called as critical activity, if the following conditions are satisfied :

1. LOT and EOT are equal at the head event.
2. LOT and EOT are equal at tail event.
3. Difference between EOT at head and tail event of the activity equals to the activity time.
4. Difference between LOT at head and tail event of the activity equals to the activity time.

Review of computation results suggests that the critical activities in the project are, A, D, E and F.

CRITICAL PATH METHOD (CPM)

Identification of Critical Path

❑ Critical path is the chain of critical activity spanning the network from start to end, i.e., the path joining all the critical events.

❑ It is also the longest path from start to end of the project network.

❑ Comparing all the possible path lengths can identify the critical path (see flow diagram).

❑ Critical path time is the shortest duration of the project.

❑ The critical path is denoted by denoting the critical events on the path. Critical path for the project is A-D-E-F.

❑ Critical path can also be denoted in terms of event numbers. In this project : 4-4-5-6

❑ That a project has multiple critical paths. In such case the length of all the critical paths will be to distinguish the critical path from other paths, use a thicker line to demarcate the critical path.

❑ It is quite possible equal.

Critical Path and Project Management

The critical path time being the shortest project time, any delay in completion of any of activity on the critical path would delay the entire project. Therefore, it is the critical activity that needs to be monitored for timely completion of project.

Conclusion

Marketing system in Case-II is more efficient than that of Case-I, because price spread is less in Case-II than Case-I. The producer's share in consumers rupee is more in Case-II than that of Case-I. These are because of less number of intermediaries involved in Case-II than Case-I. So, the marketing efficiency will be more where the intermediaries are minimum in the marketing system.

To estimate the rationality of a loan the economic feasibility analysis of credit proposals are required. The tests are conducted for:

(*a*) Returns from the proposed investment

(*b*) Repaying capacity (it will generate)

(*c*) Risk bearing ability of the borrower

The above three are popularly known as the 3 R's of credit.

Questions to be examined

1. Will the investment produce sufficient returns to cover the principal and additional costs?

2. Will the borrower have sufficient repayment capacity to return the loan and interest on it when these amount are due?

3. Does the borrower have the capacity to meet the risk and uncertainties involved in using the borrowed fund?

   (*a*) Returns to Investment.

   (*b*) Repayment Capacity.

   Part of annual income earned by a farm family available for the repayment of loan.

   (*c*) Risk bearing ability.

Marginal Analysis

Revenue or Income side

Additional income over existing plan.

Expenditure

Interest at the rate of 18% for Rs. 41,250.

Other expenses

Total expenses

Net Marginal Return (NMR)

Total loan

For Rs. the net marginal return is Rs.

Percentage of marginal return

The percentage of net marginal return is more than the opportunity cost. Hence loan can be sanctioned.

Repayment Capacity

Repayment Capacity for self Liquidating loan = Gross income – (living expenses + working expenses excluding proposed loan + taxes + other loans + repayment dues)

Non-liquidating loan = Gross income – (living expenses + working expenses including seasonal loans + taxes + other loans + repayment dues)

	Without credit	*With credit*
Gross return		
Operating cost		
Net return		
Living expenses		
Repayment fund/others		

Total repayment capacity

For the loan of Rs.? The interest at the rate of 18 per cent become Rs.?

Repayment per year

Because the repayment/year of the borrower is less than his repayment capacity, the loan can be sanctioned.

Risk Bearing Ability

Coefficient of variation

Gross returns without credit

Gross return with credit

Repayment capacity for non-liquidating loan (proposed)

The repayment capacity at risk is more than the repayment due per year. So the loan may be sanctioned.

Objective : Setting up a small acquarium enterprise for rural women.

Reasons for starting up a enterprise

1. To utilize the free time women hour.
2. So that they can develop creativity and interest in work.
3. As a supplement to finally income.
4. To stand on their own feet.
5. The generate employment for others also.

Why select aquarium as a business?

1. Raw material cheap and easily available.
2. Not very much time consuming.
3. Can be taken as a part-time job
4. Can be started at a little space with less inputs.
5. Can be expanded to maximize outputs.

How to Start

Location

The aquariam can either be made at home if space is available or take a rented room in the main market. The room can be minimum of 5 × 6 m.

Finance

A heavy finance is not required to start the business but still the required money can be arranged from any local banks. A minimum of Rs. 20,000/- loan can be taken.

Raw Material

The raw material can be purchased from wholesale shops at a cheap rate. In the starting we don't use very expressive material.

❏ We taken plain glasses according to size
❏ Filters
❏ Low quality aerator
❏ Thermostate
❏ Sealant
❏ Stones, sand, gravels
❏ Scenary
❏ Artificial plants.
❏ Locally available fishes.

The Requirements of the Room

The room chosen for the business purpose must fulfil the following requirements.

❏ Divide the room into 3 parts.
❏ Keep the front area for display of model aquarium.
❏ For storage a shelf and cupboard is required where we can keep glasses, filters etc.
❏ In the centre place a table for making aquariums.
❏ The room must have tap attached for running water supply.
❏ Keep the room clean and attractive.

Pricing

It will be determined by adding raw material cost, labour cost, transportation cost and profit.

Advertising

❏ Newspaper
❏ Pamphlets
❏ Advertisement on local channels.
❏ Exhibitions

PROSPERITY FOR ALL AQUARIUM FOR ALL

Introduction : The ultimate aim of education is to raise the standard & quality of life. The entrepreneurship helps us in this direction. Our Project is based on aquarium business. The aquarium management is newly growing and hot enterprises which is gaining momentum in town area besides from long-term utilization in metro cities. The enterprise will be a small scale enterprise with a small sufficient working capital. It will be run in rented shop for starting year and may get shifted depending upon the market and consumer response. The market is having enough potential to accept new enterprise as observed by market survey. The aquarium is not only a means of physical decoration but also serve as a means of religious satisfaction to many peoples as to see fish in the morning is believed auspicious by many people, hence forth, people are attracted to have aquarium in home, the religious importance of fish has also supported by Chinese culture through feng shui which is now well accepted in Indian home also.

Content of Project

1. General Information

(a) **Biodata**

S.No	Name	Age	Sex	Qualification	Experience	Capability

(b) **Industrial Profile :** The proposed enterprise will be a newly set enterprise of its own kind with a hand of 6 month entrepreneurs.

(c) **Constitution & Organization :** The entrepreneur will be of partnership kind whereby all the enterprise related work will be clearly mentioned practiced also the equal Profit share clause be and followed.

Self Product Details

1. Utility of Product : The proposed product can serve both as a means of physical decoration and also as a mean of religious of satisfaction. It will be a perfect model of an enclosed artificial alive habitat for small children which will

- ❏ Decorative Utility
- ❏ Religious Utility
- ❏ Education Utility
- ❏ *Emotional Utility :* People who are living alone can pass their time constructively and creatively through decorative fishers.
- ❏ *Decorative Utility :* The Aquarium can emphasize the centre of attraction at any corner of house.

Product Range: The Aquarium has an additional quality of availability in different forms & shapes: The size of Aquarium can be increased or decreased, altered, moulded etc. as by the taste and need of customers.

The different size & shape of Aquarium have different price value depending upon their decorative and potential capacity to maintain the fishes.

Product Design/Drawing

1. Rectangle Form

2. Square Form

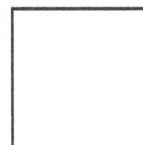

3. Hexagonal Form

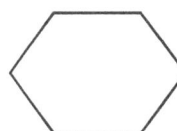

4. Spherical Form

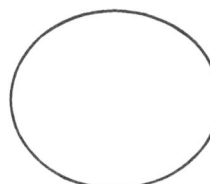

S.No	Size	Shape/form	Price
1.	12 × 24 × 12	Rectangle	2500
2.	12 × 48 × 12	Rectangle	3000
3.	60″	Rectangle	5000
4.	24″	Square	2000
5.	48″	Square	2500
6.	60″	Square	3000
7.	24″	Hexagonal	3000
8.	48″	Hexagonal	3500
9.	60″	Hexagonal	4000
10.	24 (diameter)	Spherical	

Specially moulded in different shapes are also available

(a) Swan shaped

(b) Rah shaped

Employment generation : The enterprise is capable to provide employment to 2 persons, the day it will be started. It will be provide continuous & regular employment opportunities for the needful and will increase in coming year according to the demand.

1. Employee for maintenance the initial amount will be given as 1,000 Rs/month to one employee who will be maintaining aquarium.

2. For transportation: the second employee will shift the sold aquarium to the customer house and helps in installation and placement.

Project Description

1. **Site:** The project will be developed at main market of haldwani at kaladhungi chauaraha or any city mol. The site land will be leased and these will be a two room show room of aquarium in which one room will be used as construction and one will be as show room. The site is suitable because in main market the people can see it frequently and can attract toward the aquarium.

2. **Physical Infrastructure:**

   (a) *Raw material :* The raw material like glass, silicon, rubber, pabbels, fishes are available in the near by market. These in no need of any import of any raw material.

   (b) *Skilled labour :* As it requires skills like glass cutting, fixing of glass and maintenance of the aquarium. So that type of labour is easily available in the near by surrounding.

   (c) *Utility :* 1. *Power :* for this purpose not more power supply is needed but the normal power supply is needed for this.

   - *Fuel* : No any fuel is required for that purpose

   - *Water* : Water is the essential part for the aquarium business. So the adequate water supply is needed. The type water is enough for that purpose.

3. **Pollution control :** As it is not an effluent releasing ventured. There is no chance of pollution. So it is pollution free venture.

4. **Communication system :** For the purpose of communication, these will be 2 phones (bases) and email id of the owner, mobile phone.

5. **Transport facility :** As the name material is available within the market so for the transport we can hire auto jeep, rickshaw.

6. **Manufacturing process**

   (a) *Material required*

   1. Glass – 24″ × 12″ – 3 glass

       12″ × 12 – 2 glass

   2. Silicon rubber – 1 tube

   (b) *Procedure:*

   1. Cuts the glass pieces according to the size.

   2. Clean the glass pieces neatly.

   3. Adjust glass in such a way that there is no space for water leakage and mark it.

   4. Apply the silicon rubber with the help of the spatula on one base of glass which is to be joined and so on the glass is to be placed and make rectangular aquarium.

   5. After 10-15 minute again sealing is done with the help of silicon rubber to seal any space if left.

   (c) *Precaution:*

   1. Adjust the glass very well.

   2. Do the mark with cage otherwise the edges of glass can damage any part of the body.

3. Sealing should be done carefully.

4. Glasses should be free from air bubbles.

5. Places the glasses as soon as silicon is applied on it otherwise it will dry.

6. Cover silicon rubber with polythene, uncovered sealant should not come into contact with the ice or moth.

7. Prologue contact of sealant with the skin should be avoided.

7. **List of Machinery & Equipment**

Equipment	Price
Glass	250/aquarium
Sealant	125/do
Aerator	150
Thermostate	300
Filter	200
Pebbles	10/kg
Feed	120/kg
Feed rings	10
Decorative plods	100

8. **Technology selected:** The technology is update because all the work has to be done by manually so no additional mechanical technology is needed.

9. **Research and development:** For research and development we collaborate with the college of fishery science of G.B. P.U.A&T Pantnagar to know new researches about the aquarium management and it's precaution.

(A) Equipment Already Purchased

Item	Rate	No's	Value in Rs.
Glass	250/aquarium	50	2500/-
Sealent	125	4	600/-
Aerator	150	10	1500/-
Thermostate	300	10	3000/-
Filter	200	10	2000/-
Fishes	30/-	100	3000/-
Pebbles	100/-	10kg	1000/-

(B) Equipment to be purchased

1.	Feed	120/kg	5kg	600/-
2.	Feed rings	10/-	20Nos.	200/-
3.	Decorative item & artificial Plant	100/-	20Nos	2000/-
	Total cost =		A + B	= 27500.00

(C) Details of Preliminary expenses

S.No		Rs.
1.	Advance for Shop rent	15000/-
2.	Advance for electricity undertaking	2000/-

3.	Advertisement	1000/-
4.	Furniture	10,000/-
5.	Lighting	2000/-
6.	Water connection	1000/-

(D) Cost of Production & Profitability Estimates

		1	2	3	4	5	6	7	8	9	10
*	Installed capacity	10	15	20	30	30	35	-	-	-	-
*	Capacity									Utilization	
	(% of capacity)	100	100	100	100	100	100	100	-	-	-
*	Labour	2	2	3	3	4	4	4	-	-	-
*	Wages & salary 500/month										
	Total	1200	1200	1800	2400	-	-	-	-	-	-
	Total Cost of Production = A + B + C + D										

Capital Cost and Sources of Finance

1. 1-Shop rent
2. 2-Furniture
3. Tubes
4. Bulbes
5. Curtains
6. Vehicles
7. Tools
8. Fables

Contingency Fund against price rise unexpenses.

Sources of Finance

Bank + Self deposited.

Assessment of working capital requirement:

1. Bleak even paint

2. Return in investment

Employment generation : The entrepreneur will employ 2 workers initially by will increase the no of employees with expand of business.

Local Resource Utilization

Local fishes can be used at a low cost which will add support to our business.

Development of area: As we will set up our business in haldwani city mol the shop will be unique as the awness will be expertise in the field.

AQUARIUM BUSINESS PLAN

Aquarium business requires not much investment and it can give high returns if planned and managed well. Initially, the business can be started in a room having racks and shelves to hold the equipments and decorative material *e.g.* filter, bubblers, artificial plants of different colours and sizes, shells, scenery or the film which is pasted on 1 side of the Aquarium there should be some source of fresh water, pipes to connect the filter and bubbler. The basic material to construct an aquarium *i.e.,* glasses and silicon rubber should also be available. Some sample fishes like goldfish, guppy, white shark can be kept in an aquarium for display.

After the business starts flourishing a bigger area can be selected where a no. of Aquarium of different shapes and designs can be exhibited, this will make the selection easier for customers. Different designs of stands and covers can also be displayed.

More variety of fishes can be displayed. Filters, bubblers and feed can be manufactured if the business gets popularity.

In order to popularize the business, advertising and customer dealing should be effective.

For advertising, pamphlets can be distributed defining the special features or Advertisements in newspaper can be given.

The interior of the shop should be attractive and lively so that it is appealing for the customers. Colourful fishes, gravel and different designs of equipments and sceneries should be selected to attract more and more customers. The place should be neat and clean and well-organised. All the material should be placed in its place for easy view. There should be glass racks so that customers can view and select.

Economics of Aq. shop at Small Scale

Rent of Shop	Rs. 3000/-
Glasses (for 3 Aq.)	Rs. 750/-
Silicon Rubber	Rs. 100/-
Covers (3)	Rs. 300/-
Stand (3)	Rs. 450/-
Fishes (25.30)	Rs. 450/-
Gravel	Rs. 50/-
Artificial Plants	Rs. 100/-
Lights	Rs. 500/-
Equipments	Rs. 1000/-
Miscellaneous	Rs. 2000/-
	Rs. 8700/- ; Rs. 9000/-

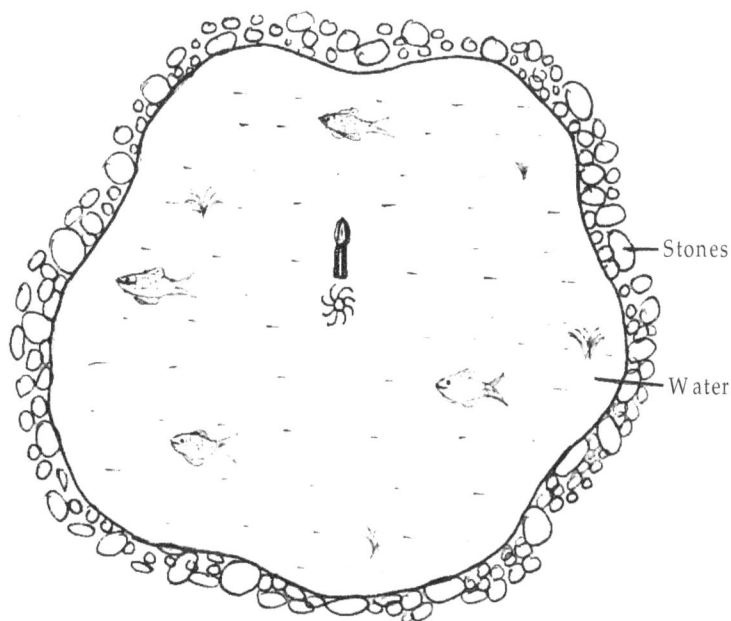

POND

INITIAL PLAN OF BUSINESS ACTIVITY

Measurement of Room
5 mt. × 6 mt.

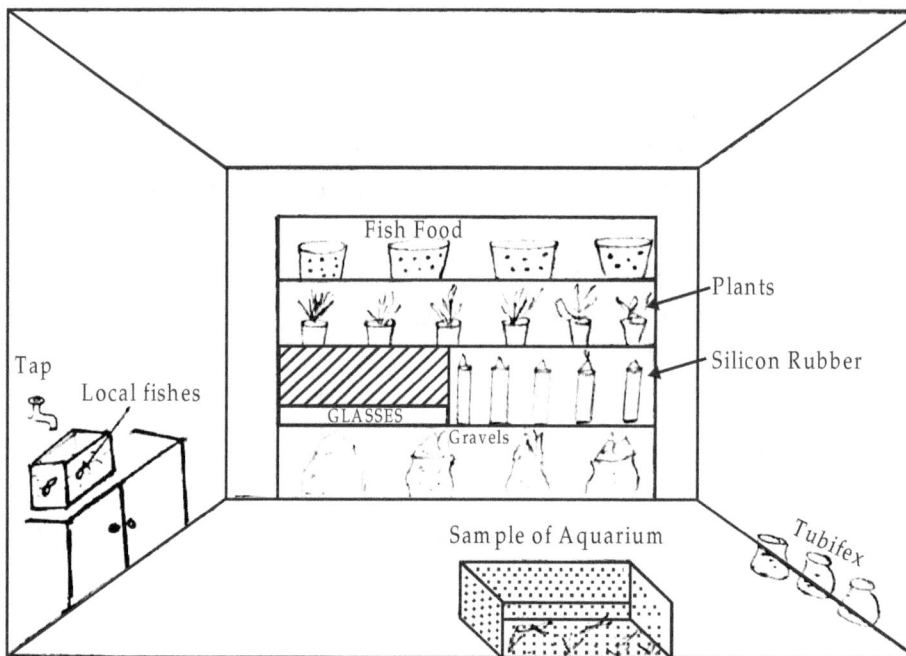

Fish Food

Plants

Silicon Rubber

Tap

Local fishes

GLASSES

Gravels

Sample of Aquarium

Tubifex

EXPANDED BUSINESS PLAN

Measurement of Room
10 mt × 10 mt.

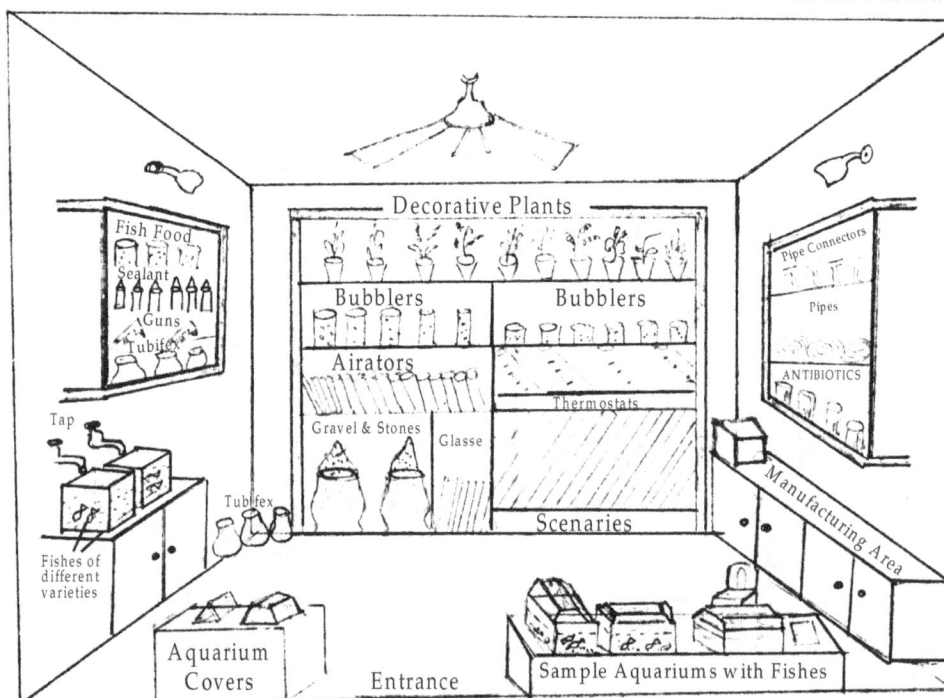

Decorative Plants

Fish Food

Sealant

Guns

Tubifex

Pipe Connectors

Pipes

ANTIBIOTICS

Bubblers

Bubblers

Airators

Thermostats

Tap

Gravel & Stones

Glasse

Tubifex

Scenaries

Manufacturing Area

Fishes of
different
varieties

Aquarium
Covers

Entrance

Sample Aquariums with Fishes

Introduction

Ornamental fish keeping is one of the most popular hobbies in the world today. The growing interest in aquarium fishes has resulted in steady increase in aquarium fish trade globally. The trade with a turnover to US $ 5 billion and an annual growth rate of 8% offers a let of scope for development.

The top exporting country is Singapur followed by Hongkong, Malaysia, Thailand, Philippines, Sri Lanka, Taiwan, Indonesia and India. The largest importer of ornamental fish is the USA followed by Europe and Japan. The emerging markets are China and South Africa. Over US $ 500 million worth of ornamental fish are imported into the USA each year.

India's share in ornamental fish trade is estimated to be Rs. 158.23 lakh which is only 0.008% of the global trade. The major part of the export trade is based on wild collection. There is very good domestic market too, which is mainly based on domestically bred exotic species. The overall domestic trade in this field cross 10 crores and is growing at the rate of 20% annually.

The earning potential of this sector has hardly been understood and the same is not being exploited in a technology driven manner. Considering the relatively simple technique involved, this activity has the potential to create substantial job opportunities, besides helping export earnings.

Why Breeding?

Ninety five percent of our ornamental fish export is based on wild collection. Majority of the indigenous ornamental fish trade in India is from the North Eastern states and the rest is from southern states which are the hot spots of fish biodiversity in India. This capture based export is not sustainable and it is a matter of concern for the industry.

In order to sustain the growth it is absolutely necessary to shift the focus from capture to culture based development. Moreover most of the fish species grown from their ornamental importance can be bred in India successfully. Organised trade in ornamental fish depends on assured and adequate supply of demand, which is possibly only by mass breeding.

Technology

There are quite a large number of tropical aquarium fishes known to aquarists. While many of the fish are easy to breed. Some of there are rare, different to breed and expensive. Most of the exotic species can be bred and reared easily since the technology is simple and well developed.

It is advisable to start with common, attractive, easily bred and less expensive species before attempting the more challenging ones. An ornamental fish project can be either.

* Rearing only
* Breeding only
* Breeding and rearing depending upon the space available.

The technology involves the following activities :

Culture/Rearing : The culture/rearing of there fishes can be takes up normally in cement tanks. Cement tanks are easy to maintain and durable.

One species can be stocked in one tank. However, in case of compatible species two or three species can occupy the same tank. Ground water from dug wells/deep tubewells are the best for rearing. The fishes reach the marketable size in around 4 to 6 months. Eight to ten crops can be taken in a year.

Feeding : Young Fish are Fed mainly with Infusoria, Artemia, Daphnia, Mosquito larval, Tubifex and blood worms. For rearing, formulated artificial or prepared feed can be used. At present no Indigeneous prepared feed for aquarium fish is available.

The amount and type of food to be given depends on the size of the fry. Feeding is generally done twice in a day or according to requirement. For rearing from fry stage dry/ prepared feed can be used.

Breeding : The method of breeding is based on the family characteristics of the fish. The success of breeding depend on the compatibility of the fish pair, the identification of which is a skill born out of experience.

Generally the breeders are selected from the standing crop or purchased and reared separately by feeding them with good live food. However, it is always better to buy good breeding stock. Otherwise, the original characteristic of the species keeps on getting diluted because of continuous inbreeding.

Breeders especially egg layers should be discarded after few spawnings. Health care, Water exchange, is a must for maintaining water quality conducive for the fish health. Only healthy fish can with stand transportation and fetch good price.

Chemicals/Antibiotics, Methylene blue, Methylene yellow, Malachite green, Ampicilin, Vitamins, Potassium permanganate, Copper sulphate etc. can also be used, for preventing/ treating diseases.

Market : At present the market is mainly domestic. There is a good domestic market which is increasing. The export market for indigenously bred exotic species is also increasing.

Ornamental Fishes : Aquarium Fishes are mainly grouped into two categories, *viz.,* Oviparous (egg layers) and Viviparous (live bearer). Further, the Fresh water ornamental Fish varieties can be broadly grouped into Tropical and cold water species also. Management of there two categories are different in nature. According to water tolerance Fishes are hard water tolerant and soft water tolerant species and those wide tolerance. The Fishes and the details of grouping is given below :

Species	Water Quality	Breeding Season	Breeding Type	Egg Type/Care
Rosy burb	Wide Tolerance	Summer/Mansoon	Egg scatterer	Adhesive
Gold fish	Wide Tolerance	Mansoon/Winter	Egg scatterer	Adhesive
S Fighter	Wide Tolerance	Summer/Mansoon	Nest builder	Male Guard Eggs.
Cat fish	Wide Tolerance	Mansoon/Winter	Egg depositer	Encloser Regd.
Angel	Soft Water	Summer/Mansoon	Egg depositer	Parents Fan Eggs.
Cichlid	Soft Water	Summer/Mansoon	Egg depositer	Enclosed Regd.
Manila Carp	Soft Water	Summer/Mansoon	Egg scatterer	Adhesive
Molly	Hard Water Species	Summer/Mansoon	Live bearer	Young ones
Guppy	Hard Water Species	Summer/Mansoon	Live bearer	Young ones
Platy	Hard Water Species	Summer/Mansoon	Live bearer	Young ones
Sword Tail	Hard Water Species	Summer/Mansoon	Live bearer	Young ones
Blue Gourami	Wide Torelance	Summer/Mansoon	Nest builder	Male Guard Eggs.
Pearl Gourami	Wide Torelance	Summer/Mansoon	Nest builder	Male Guard Eggs.

Ornamental Fish Breeding Project

The basic requirements for successful breeding and rearing of ornamental Fish are adequate space, quality water and sufficient feed considering thin the following investment are required for starting an ornamental fish project.

Tanks

The tanks can be of RCC or brick masonry work having flat bottoms with inlet and outlet pipes, Clay, Cement, Fibre glass, or plastic tanks can also be used. Rearing of fish should be done is large tanks. Size of the tanks vary according to the space, the number and type of fish culture.

Aquariums

Glass tanks of varying size are required for breeding. Small glass bottles of 250 ml are used for keeping individual male figure fish. Number and size of the glass tanks depend on the specific breeding/spawning behaviour of the species selected.

Overhead Tanks

An overhead tanks of suitable size for storing and to enable sedimentation of water is required.

Water Supply

Deep tubewells would be the best source of water. Recycling of water through bio-filters or other sort of filtering mechanism can be tried. Other sources like dug wells, Municipal water if available can also be used. A small pump to lift the water to over head tank and a network of pipes are needed to feed the culture tanks.

Work Shed

Work Shed should be designed in such a way that the tanks get filtered sunlight. Translucent HDPE Sheets can be used dropping etc.

Aeration Equipments

A blower pump with network of tubes for aeration is a must. Continuous power supply should also be ensured through generator set or UPS or inverter.

Finance Viability

Considering the seasonality in breeding and consequent availability of seed material for stocking, the operations are going to be seasonal. In order to best utilize the installed capacity. It may be necessary to combine operations by breeding a main species that may be breeding in winter. The fecundity and the no. of spawnings assumed of some of the popular species are given below :

Species	Average Fecundity	Spawning/Year
Molly/Guppy/Sword Tail	130	12 (gives offsprings in batches)
Blue Gourami	3500	10
Pearl Gourami	800	10
Rosy Barb	700	10
Tiger Barb	500	10
Zebra/Pearl/Veiltail Danio	1000	10
Angel	800	12
Serpae Tetra	800	10
Gold Fish	3000	3

Assumption

For the purpose of working out economics of breeding unit, a unit size 300 sq. mt. with tank volume of 60 m^3 has been considered with an average production capacity of 2 lakh fry per year. A combined operation of summer/mansoon species and winter species has been considered for working out the economics in this model. Unit of this size has been designed considering the small entrepreneurs in view.

However, the same could be increased on modular basis and the economics can be worked out in project situation accordingly. The larval rearing has been assumed as 40 days. The recurring cost has been assumed with a fecundity of 800 and a survival of 50% upto 40 days. The breeding percentage is taken as 60 per cent.

The no. of brood fish depends on the Fecundity and survival of each species so as to get the combined annual production of about 2.5 lakh fry at the end of 40 days. Only 50% production is assumed for the first year. The sale price is assumed Re. 1.00 per fry for mansoon species and Rs. 2.50 per fry for gold fish.

The financial analysis has been shown in Annexure III. The result of the analysis are :

NPW at 15%   DF : Rs. 244436

BCR at 15%   DF : Rs. 1.36 : 1

IRR is 35%

Margin Money and Bank Loan

The entrepreneur is expected to bring margin money out of his own resources. The rates of margin money stipulated are 5% for small farmers 10% for medium farmer and 15% for other farmers. For corporate burrowers the margin stipulated is 25%.

NABARD could considered providing margin money loan assistance is deserving cases.

Rate of Refinance

NABARD provides refinance assistance for ornamental fish rearing to commercial banks, Cooperative banks and Regional Rural Banks. The rate of refinance is fixed by NABARD from time-to-time.

Interest Rate for Ultimate Borrowers

Banks are Free to decide the rate of interest with in the overall RBI guidelines. However, for working out the financial viability and bankability of the model project we have assumed the rate of interest as 12% per annum.

Interest Rate for Refinance from NABARD

As per the policy circulars of NABARD issued from time-to-time.

Repayment Period

The borrower will be able to repay the bank loan in 6 years with a grace period of one year on repayment of the principal.

Security

Banks may take a decision as per RBI guidelines.

Subsidy

The MPEDA provides subsidy at the rate of 50 per cent of capital cost towards construction of cisterns, glass aquarium tanks, aeration system, oxygen cylinder, electrical fittings and essential accessories excluding cost of construction of shed, subject to a maximum Rs. 40000 per unit.

Some State govt. also provide subsidy through their departments.

Model Project for Ornamental Fish Breeding Unit

A model of ornamental fish seed hatchery is given below. The parameters are averaged out and costs are only illustrative. The cost of different parameters change depending on the area, the type of tanks and the species bred. According to the place and requirement of the project, some of the items can be excluded or more items as required may be included.

Annexure I : Project Cost for Ornamental Fish Hatchery

Hatchery tank area (sq. m.)	100
Hatchery total area (sq. m.)	330

Item of Investment	Quantity	Rate	Total Cost
A. Building and Civil work			
1. Hatchery shed with A/C sheet roofing & side wall	330 m^2	1200	369000
2. Tank vel (Ltd.)	60000	1.25	75000
3. Flooring (m^2)	200	10.00	2000
4. Drainage pit and network			12000
5. Water supply network			3000
6. Filteration system/Outlet			3000
7. Electrification and Installation			5000
B. Machinery and Equipment			
1. Air blower (3 hp × 1 no.)			12000
2. DG set (8 hp with 6/8 KVA) alternator			36000
3. Heater			3000
4. Sand filter			2000

(Contd...)

	Item of Investment	Quantity	Rate	Total Cost
5.	Pump (3 HP)			17000
6.	Tube well			40000
7.	Pump house			5000

C. Misc. Fixed Assets

1.	Plastic pools			5000
2.	Glass aquarium			3000
3.	Lab instruments			2000
4.	Glass wares			2000
5.	Furniture			2000

D. Preliminary Recurring Cost

Cost of brooders			5000
Feed			10000
Medicines			500
Electricity			3000
Miscellaneous			2000
Labour			10000
Total cost			*676000*

E. Production

Summer/Mansoon species (1500000 fry)		150000
Gold Fish (50000 fry)		125000
Total Income per year		*2750000*

Annexure II : Financial Analysis

A. Cost 1st Year

1.	Fixed Costs	645500
2.	Recurring Cost	3050030500
	Total	*67600030500*

B. Benefits

1.	Income from sale of fish	137500275000
2.	Net income	53850024500

Analysis

1.	Net Present worth of costs	676731
2.	Net Present worth of benefits	921168
3.	Net present worth	244436
4.	BCR – 1036 : 1	
5.	IRR – 35%	

Annexure III : Estimated Bank Loan and Repayment Period

Total Outlay	676000
Margin 25%	33800
Bank Loan	642200 (Amount in Rs.)
Interest @ 12% pa	

Year	Bank Loan	Net Income	Repayment	Net Surplus	Principal	Total
1	642200	137500	77064	—	77064	60436
2	642200	244500	77064	128440	205504	38996
3	513760	244500	61651	128440	190091	54409
4	385320	244500	46238	128440	174678	69822
5	256880	244500	30826	128440	159266	85234
6	128440	244500	15413	128440	143853	100647

Besides the cost factors, the report should includes present probable sources of finance. The sources of funds should equal the cost of a project as otherwise the project cannot be set up in full. The resources would include the owner's funds together with loans and deposits raised as well as the limits expected from financial institutions/banks.

The estimation of funds for the cost factors involved should be realistic and correct. Many units run into serious financial problems because of inadequate estimate of funds requirements.

Assessment of Working Capital Requirements

Planning for working capital requirement is equally crucial for an entrepreneur. While estimating the capital costs, margin for working capital was taken into account. Any unit will be able to function only when adequate working capital funds are available. In other words, at the initial stage itself, the estimate of working capital requirements should be made and shown alongwith the total cost of the project.

Sometime back formats for working capital assessment had been designed for limits up to Rs. 50,000 for limits between Rs. 50,000 and Rs. 2 lakhs and above Rs. 2 lakhs. As such it the entrepreneurs present their estimates in those prescribed formats, it will save time and energy for them as well as for the banker.

It has been generally noticed that the entrepreneurs present the working capital requirements is their own way which is ultimately recasted by the banker. This wastes time and creates problems. Hence, if they project their requirement in the prescribed way, it will minimise objections from the banker's side.

Other Financial Aspects

One of the objectives of setting up a project is to earn a livelihood. Besides the project set up must be able to retrieve the investments made within its life cycle. This would be possible only if the products taken up for production are adequately profitable. This would require preparation of a projected profit and loss A/c which would indicate likely Sales Revenue, Cost of Production, Allied cost and Profit. These estimates, especially the likely Sales Revenue, should be made on a realistic basis. A projeced Balance Sheet and Cash Flow Statements (as per Annex 4 & 5) would also have to be prepared to indicate financial position and requirements at various stages of the project. After all the smooth functioning of the unit necessitates availability of adequate funds for various commitments.

Next the Break-even Analyses must be presented. Break-even point is the level of production/sales where the industrial enterprise shall make no profit no loss, it will just break even, this facilitates knowing the gestation period and the likely moratorium required for repayment of loans.

$$\text{Break-even point (BEP)} = \frac{F}{S - V} \times 100$$

where   F = Fixed Costs

                      S = Sales Projected

                      V = Variables Costs

There Break-even point thus calculated will show at what percentage of projected sales the unit will break-even. (Also see Annex 2).

It is also a good idea to calculate and indicate the following ratio :

1. Profitability Ratio $= \dfrac{\text{Net Profits}}{\text{Sales}} \times 100$

2. Return on Investment $= \dfrac{\text{Net Profits}}{\text{Capital Employed}} \times 100$

3. Debt. Equity Ratio $= \dfrac{\text{Debt}}{\text{Equity}} \times 100$

4. Debt Service Coverage Ratio $= \dfrac{\text{Net profit after tax} + \text{Depreciation} + \text{Interest for one year}}{\text{Instalments} + \text{Interest (for one year)}}$

OTHER DETAILS

Project Implementation Schedule

Preferably a PERT/CPM chart can be appended to the project report. If this is not feasible then in a tabular form likely dates of completion of the following activities can be simply mentioned :

- ❑ Acquisition of land
- ❑ Registration of the Unit
- ❑ Bank Loans
- ❑ Construction of Building
- ❑ Power connection
- ❑ Ordering Plant/Machinery
- ❑ Supply of Plant/Machinery
- ❑ Installation of Plant/Machinery
- ❑ Recruitment of Workers
- ❑ Training of Workers
- ❑ Ordering raw materials
- ❑ Procurement of raw materials
- ❑ Trial Run
- ❑ Commercial Production

Plant Layout

If possible, a copy of the plant layout can also be furnished in the project report. This will assist determining sufficiency of area for present and future expansion requirements.

ANNEXURES

List of Appendices of a Project Report

The following is an illustrative list of appendices, which may be attached alongwith a project report and should be preceded by the descriptive presentation as discussed in earlier pages :

1. Equipment required.
2. Details of preliminary expenses.
3. Estimates Cost of production and profitability.
4. Labour and staff requirements.
5. Break-even point.
6. Working Capital requirements.
7. Debt survive coverage ratio.
8. Cash flow statement.

Details of Equipment Required

Sl. No.	Item	Rate (Rs.)	Nos.	Value in (Rs.)
1.	Equipment already purchased			
2.				
3.				
4.				
5.				
				(A)
				....
				....
	Equipment to be purchased			
6.				
7.				
8.				
9.				
10.				
				(B)
				....
	Total cost of requirement (A + B) =			

ANNEXURE 2

Details of Preliminary Expenses

Sl. No.	Item	Rs.
Advance for factory shed		
Advance for Electricity Undertaking		
Construction of godown		
Sales Tax registration charges		
Deposit of O.Y.T. Telephone		
Research and Development Expenditure		....
		....

Cost of Production and Profitability Estimates

	Operating years									
	1	2	3	4	5	6	7	8	9	10
Installed capacity										
Production (quantity)										
Capacity Utilisation (% or Installed capacity)										
Raw Materials and Consumable stores										
Raw Material										
Dyes and chemicals										
Consumable stores packing Material										
A. Total										
Material Cost										
Utilities										
Power										
Fuel										
B. Labour										
Wages and Salaries										
C. Total Labour										
Factory Overheads										
Repairs and Maintenance										
Rent, rates, taxes and other										
Miscellaneous expenses										
Depreciation										
D. Total Factory Overheads										
E. Total Cost of Production (A + B + C + D)										

Administrative Expenses

Salaries Including Provident Fund, Light, Postage, Telegrams etc. Management Remuneration

F. Total Administrative Expenses

Contd...

	1	2	3	4	5	6	7	8	9	10
G. Selling Expenses										
H. Total Cost (E + F + G)										
I. Total expected sales										
J. Profit Before Interest (I – H)										
Financial Expenses										
Interest on term loans										
Interest on bank borrowings										
K. Total Financial Expenses										
L. Operating Profit (J – K)										
M. Other Income										
N. Profit Before Tax (L – M)										
O. Provision for Taxation										
P. Profit after Tax (N – O)										
Less : Profit to be withdrawn or										
Dividend Proposed on Equity										
Q. Retained Profit										

Break-Even Analysis

Sales Realisation	*Rs.*
(A) Total Fixed Cost	Rs.
(B) Contribution (Sales realisation – Total variable cost)	Rs.

(C) $\text{Contribution per unit} = \dfrac{\text{Total fixed cost (A)}}{\text{No. of units to be produce}}$

(D) $\text{Break-even point in terms of no. of units} = \dfrac{\text{Total fixed cost (A)}}{\text{Contribution per unit (C)}}$

(E) $\text{Break-even in terms of plant capacity} = \dfrac{\text{Total plant capacity}}{\text{Break-even production (D)}}$

Working Capital Requirements

(A) Anticipated monthly sales	Rs.
(B) Cost of production per month	Rs.
(C) Cost of interest per months	Rs.

ANNEXTURE – 1

Item	Stocking/payment period required	Working Capital Value (Rs.)	Margin (Rs.)	Permissible limit (Rs.)
(i) Imported raw material	months			
(ii) Indigeneous raw materials	months			
(iii) stock-in-process	weeks			
(iv) Finished goods	weeks			
(v) Sundry debtors	months			
(vi) Expenses for	Months			

Total				
Total working capital required		Rs.	(A)	
Less				
(i) Liquid Surplus in balance sheet as on		Rs.		
(ii) Credit on purchases (... months)		Rs.	(B)	
(iii) Permissible limit Deficit (A – B)		Rs.	(C)	
How is this to be net ?				

Cash Flow Statement

(Rs. in Lakhs)

	Operating years								
	Construction period	1	2	3	4	5	6	7	8

1. **Sources of Funds**

   Net profit before taxes with interest added back but after depreciation and investment allowance Rebate

2. Share capital : Equity/Preference

3. Depreciation

4. Investment Allowance Reserve

5. Increase in long-term loans/debentures

6. Increase in deferred payment facilities

7. Increase in unsecured loans and deposits

8. Increase in bank borrowings for working capital

9. Sale of assets/investments

10. Others (indicate details)

   Total (A)

Chapter 16

ECOSYSTEM STRUCTURE

Odum's theory is interesting in that although it is not strictly quantitative, it makes enough specific qualitative prediction for rigorous specific qualitative prediction for rigorous tests of its validity to be performed. Thus, food-web models constructed as described above can quantify many of the attributes of ecosystems that are part of Odum's theory (Table 1). As might be seen, this theory essentially implies that as systems mature, their biomass will tend to increase, especially the biomass of large animals with high longevities, and detritus will increasingly be recycle through a web whose complexity will tend to increase. Two direct tests of this theory have been performed so far. One consisted of forcing an increase in the biomass of top predators in two marine ecosystems, one coastal and one off shore, and using the Monte Carlo 'EcoRanger' routine to identify parameter values which were randomly selected from within the distribution assumed for each of the models' inputs and where compatible with these increased biomasses. This led to increases in all parameters related to increased maturity in Table 1, notably detritus recycling. The other test consisted of an application of Ecosim the dynamic version of Ecopath, to be presented below. Therein, a short strong increase in the fishing mortality of the small pelagic species dominant in each system was simulated and the time for the system as a whole to recover was plotted. Here again, detritus recycling was the ecosystem parameter which best correlated with the form of resilience implied here Ecopath approach can be used to operationalize Odum's theory and to test its basic tenets.

The flow charts generated by Ecopath can also serve to test Ulanowicz's theory of ascendancy as a measure of ecosystem development. This theory combines the information contents embedded in the different flows within the system with the magnitude of these flows to derive a combined measure of information and flow, expressed in 'flow bits'. Here, results were not as unequivocal as in the case of Odum's theory. Ascendancy did not correlate positively with maturity as expressed using a combination of parameters derived from Odum's theory, the implication probably being that Ulanowicz's theory is in need of revision.

Overall, these two examples illustrate that the existing wide availability of quantified food webs constructed using the Ecopath approach can be a boon to theoretical ecology, enabling the testing of hypotheses that have long remained untested and consolidating existing knowledge on the functioning of ecosystems.

Table 1 : Selection from Odum's (1969) list of 24 attributes of ecosystem maturity.

No. in Odum's list	Ecosystem attributes	Development stages	Mature stages
1	2	3	4
1.	Gross production/respiration	>1 or <1	Approaches 1
2.	Gross production/biomass	High	Low

(Contd...)

1	2	3	4
3.	Biomass supported/energy flow	Low	High
4.	Net community production	High	Low
6.	Total organic matter	Small	Large
12.	Niche specialization	Broad	Narrow
13.	Size of organisms	Small	Large
15.	Mineral cycles	Open	Closed
16.	Nutrient exchange between organisms and environment	Rapid	Slow
17.	Role of detritus in nutrient recycling	Unimportant	Important
21.	Nutrient conservation	Poor	Good
22.	Stability (resistance to perturbations)	Poor	Good

Particle-Size Spectra in Ecosystems

Using data from particle-size counters, proposed that the size spectrum of marine organisms is a conservative feature of marine ecosystems, characterized by a constant slope, for which they provided the rationale, using thermodynamic considerations. The controversy which ensued, still festering among marine biologists, who tend to consider only the small range of organism sized sampled by automatic plankton particle counters, was largely ignored by fisheries scientists, who quickly established not only that fish abundances also neatly fit (log) linear size spectra, but that the slope of such spectra reflect the exploitation level to which a multispecies fish community is subjected, being steeper were exploitation is high.

This provides another constraint for multispecies and/or ecosystem models, which should be expected to generate such spectra as one of their output (as did, the North Sea model of Andersen and Ursin already described).

Individual-based models such as OSMOSE can generate such spectra, as may perhaps be expected. Perhaps more surprisingly, it turns out, however, that mass-balance models, can also be used to generate size spectra of the ecosystem, and thus build a bridge between the hypotheses advanced by marine ecologists and those of fisheries scientists routine of Ecopath which is used to re-express the biomass of fish, invertebrates and marine mammals in various ecosystems in the form of standardized size spectra, whose slope can then be compared. This routine, state does the following :

- Uses the von Bertalanffy growth curves and the values of P/B (*i.e. Z.*) entered for each group in the model to re-express its biomass in term of a size-age distribution;
- Divides the biomass in each (log) weight class by the time, Δt, required for the organisms to grow out of that class (to obtain the average biomass present in each size class).
- Adds the $B/\Delta t$ values by (log) class, irrespective of the groups to which they belong.

Figure 2 compares size spectra obtained in this fashion for a tropical model of the Gulf of Thailand. The spectra are based on 40 ecosystem groupings ranging in size from phytoplankton to dolphins. The spectra are obtained by running a model describing the 1973 situation, where fishing pressure and resources depletion were moderate, with observed fishing pressure through to 1994, where fishing pressure was high, and the resources severely depleted, Further, a simulation was performed with fishing pressure removed, and the long-term equilibrium used to estimate a spectrum without fishing.

The dots on Fig. 2 indicate the actual size spectrum obtained for the 1973 model. It is far from a straight line, but shows a hump in the size range dominated by benthos. This hump is likely to be caused by poor resolution for the benthic groupings, and would probably not appear if there groups were split into finer taxonomic groups than used for the model presented.

The three straight lines on Fig. 2 show that the slope of the size spectra, as expected, increases with fishing pressure, here from (-0.17) for the unfished situation, to (-0.18) for the moderately fished, and to (-0.22) for the situation with high fishing pressure. The increased slope is caused by the removal of virtually all higher trophic-level organisms in the Gulf during the time span studied.

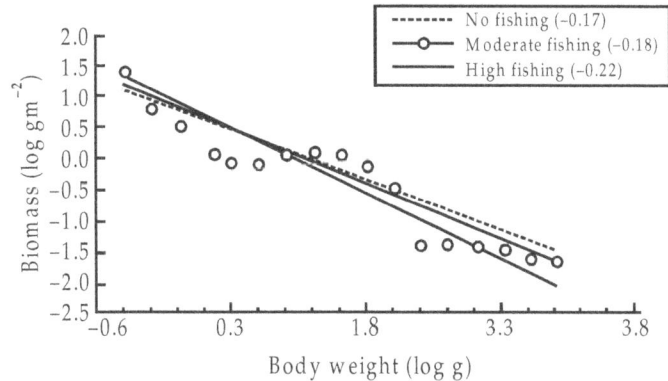

Fig. 2 : Ecosystem-level sized spectra for the Gulf of Thailand. The spectra are derived based on a trophic model with 40 ecosystem groupings (see www.ecopath.org). The thin line associated with 'moderate fishing' describes the 1973 situation in the Gulf, while the broken line indicates the slope of the model situation at long-term equilibrium without fishing, and the thick line indicates the 1994 situation, with resources severely depleted by over-fishing. Note that the slopes of the spectra (in brackets) increase with fishing pressure.

The example given here only serves as a taster for how size spectra may be of use as ecosystem indicators of exploitation level and no generalizations have been drawn so far of how the slopes of the spectra relate to exploitation. We anticipate, however, that the ease with which ecosystem-level size spectra can now be constructed through the Ecopath approach will lead to a blossoming of spectral analysis in the foreseeable future, and that valuable insights will be gained in the process.

Ecosim, Refugia and Top-Down Vs. Bottom-up Control

The system of coupled linear equations that is behind a balanced Ecopath model (see equation) above can be re-expressed in terms of the change implied by:

$$dB_i/dt = g_i \cdot \sum Q_{ij} - \sum Q_{ji} + I_i - (M_{0i} + F_i + E_i) B_i$$

where dB/dt is the rate of biomass change, g the growth efficiency (*i.e.* P/Q), F the fishing mortality (*i.e.* Y/B), M_0 the natural mortality (i.e. excluding predation), I is the immigration rate, E the emigration rate, and Q_{ij} (Q_{ji}) the consumption of type *j* (i) biomass by type i (j) organisms.

This set of coupled differential equations could be easily integrated as they are over time, thus yielding a simulation model with the help of which various scenarios resulting from changes in fishing mortality F, could be explored. This simulation model, however, would be of the Lotka-Volterra type, wherein the amount consumed of a given prey '*i*' is proportional to the products of its biomass times the biomass of its predator (s). Such 'top-down controlled' systems, however, are inherently unstable, and usually fluctuate in unrealistic fashion. As animals do not live in reaction vats, modelling predation must reflect the ability of potential prey to hide or camouflage themselves, or generally to evolve strategies that limit their exposure to predators. In Ecoism, the dynamic counterpart to Ecopath, the existence of physical or behavioural refugia is represented by prey biomass consisting of two elements: one, vulnerable to predators, the other invulnerable. It is then the rate of transfer between two partial components of prey's biomass which determine how much of the prey can be consumed by a predator. When the exchange rate is high, part of the biomass which is vulnerable is quickly replenished and hence we still have top-down control and Lotka-Volterra dynamics. On the other hand, when the replenishment of vulnerable biomass pool is set to be slow, it is essentially that slow rate which determines how much the predators can consume. We speak here of 'bottom-up control' since it is the dynamics of the prey which shape the ecosystem. Issues of top-down versus bottom-up control are addressed further in the food webs.

These different control types can be represented by replacing Q_{ij} in the above equation by:

$$Q_{ij} = V_{ij} a_{ij} \, B_i B_i / (V_{ij} + V_{ij}' + a_{ij} B_i)$$

where Q_{ij} and V_{ij}' represent rates of behavioural exchange between vulnerable and invulnerable state and a_{ij} represents the rate of effective search by predator j for prey i, i.e. the Lotka-Volterra mass-action term.

Thus, Ecoism, which allows users to change the rates of exchange, allows testing the effect of assumptions about bottom-up vs. top-down control. As it turns out, these effects are profound: pure top-down control, as it turn out these effects are profound pure top down control which is the Lotka-Volterra assumption, generates, upon the smallest shock, violent oscillations such as do not occur in real ecosystems. Conversely, under bottom-up control, depletion of one species, as, for example through fishing, tends to affect the biomass of only that species and less strongly its key prey and predators; the system as a whole generally remains unaffected even when the species in question is one of its major components (the astute reader will note this to be the unstated assumption behind single-species population dynamics). Thus, at least in terms the implementation presented here, intermediate control, which is a form of control that is neither fully top-down, nor bottom-up, is required for simulated ecosystems to behave in realistic fashion, this being a major finding obtained through Ecoism. (*www ecopath. org* may be consulted for further information on this rapidly evolving software).

Spatial Considerations in Ecosystem Modelling

As mentioned in the introduction, adding complexity to ecosystem models does not necessarily make them 'better'. Rather, to increase the usefulness of models, what is required is to identify those improvements of existing models for which the gain in new insights outweighs the added complexity and data requirements. As it turned out, the major improvements that can be added to models such as Ecopath and Ecoism are spatial considerations, capable of representing explicitly some of the refugia implied in the above Ecoism formulation.

The formulation developed for this, still based on Ecopath parameterization and its inherent mass-balance assumptions, is one wherein the ecosystem is represented by say, a 20×20 grid of cells with different suitability to the different functional groups in the system. Movement rates are assumed symmetrical in all directions around the cell, but are higher in unsuitable habitat. Their exact value is not important. The survival of various groups and their food consumption are assumed higher in suitable habitat, but they other-wise consume prey as they do in Ecopath and Ecoism, as they encounter them within a given cell.

Starting from the Ecopath baseline where functional groups are distributed evenly over suitable habitats, Ecospace simulation iterates towards a solution wherein the biomass of all functional groups is spread over a number of cells, given the predation they experience and the density of prey organisms they encounter in each cell. The distribution maps thus predicted can be compared with existing distribution maps and inferences drawn about one's understanding of the ecosystem in question and the functional group they are in. The rich patterns obtained by the application of this approach to a number of Ecopath files, thus turned into spatial models, suggest that the broad pattern of the distribution of aquatic organisms can be straightforwardly simulated.

One particularly interesting aspect of Ecospace is that it allows for explicit consideration of ecosystem effects when evaluating the potential impact of Marine Protect Areas (MPA) in a given ecosystem, thus allowing for a transition towards ecosystem-based fisheries management.

Towards a Transition from Single-Species to Ecosystem-Based Management

The requirement for ecosystem-based management derives its validity from a stark set of alternatives: either the beginning of the 21st century will see a transition towards ecosystem-based management in most areas currently exploited by major fisheries, or it will see the destruction of the ecosystems exploited by these same fisheries. Until recently, the call for ecosystem-based management could be dismissed by regulatory agencies on continuing business as usual because ecosystem management tools were not available. The sets of conceptual tools mentioned above, notably Ecopath, Ecosim and Ecospace can, however, be used to identify the key elements of management strategies that would enable fisheries to be sustained by sustaining the ecosystem in which they are embedded. The two major tools available for this are:

1. Ecosed : a routine for seeding marine protected areas (MPAs), originally covering only one cell within the spatial grid in an Ecopath/Ecospase map, then identifiying, by brute-force computations, the sequence of cells which, when added to that initial seed, will contribute most to overall benefits, either in terms of market values calculated as species caught multiplied by their price, minus cost of catching them, and/or in terms of their existence values, for example, as whale abundance to the whale-watching industry. Results of various Ecoseed runs for different systems show, in accordance with theoretical expectation, that MPAs should be large relative to exploited areas. Important here is that MPAs simulated this way consider all trophic interactions between the species included in an ecosystem model and not only the expected biomass increase of a few charismatic species.

2. An optimization approach structured around Ecosim's ability to identify the mix of fleet-specific fishing mortality which optimizes cumulative benefits over a set period, *e.g.* 30 years, under any of the following constraints:

    (a) maximizing net return to the fishery, which generally involves moderate morality on valuable target species and a small overall level of effort;

    (b) maximizing employment, which means identifying the fleet configuration which sustainably exploits the ecosystem, albeit with an effort level as high as possible. This high leads to high employment;

    (c) maximizing ecosystem maturity, by identifying the fleet configuration which maximizes, for all groups in a system, the sum of product of biomass and B/P. The latter is in accordance with Odum (1969) who predicts the highest product of B and B/P for the long-lived organisms typical of mature systems;

    (d) mandated rebuilding, wherein the fleet configuration is sought which enables faster rebuilding towards a set threshold as required. This could result from a consequence of a legal decision;

    (e) optimizing a mix of (a)-(d) wherein any of these can be given different weights.

The two approaches combined and used independently any ecosystem to explore the implications of various policies, and to quantify their outcomes in economic and social terms, and in the form of the ecosystem biodiversity.

The concepts and tools for ecosystem modelling presented above allow for a transition towards ecosystem-based management of fisheries. Obviously, uncertainty remains a major issue in such analyses. For example, the science of how whole ecosystems respond to changes in management remains very weak. However, not using tools such as described here will not reduce uncertainty. The way to reduce uncertainty about ecosystem function and the impacts of fishing thereon is to construct representations of these ecosystems and to probe their behaviour by posing intelligent questions about the process, and intelligently interpreting the answers.

REFERENCES

Allen, K.R. (1971) Relation between production and biomass. *Journal of the Fisheries Research Board of Canada* 28, 1573-81.

Alward, G.L. (1932) *The Sea Fisheries of Great Britain and Ireland*, Grimsby: Albert Gait.

Andersen K.P. and Ursin, E. (1977) A multispecies extension to the Beverton and Holt theory of fishing, with accounts of phosphorus circulation and primary production. *Meddelelser fra Danmarks Fishkeri og Havundersogelser*, N.S.7, 319-435.

Beverton, R.J.H. and Holt, S.J. (1957) On the dynamics of exploited fish populations. United Kingdom Ministry of Agriculture and food. Fishery Investigations Series II, vol, xix, 533 p.

Bianchi, G., Gislason, H., Graham, K., Hill, L., Jin, X., Koranteng, K., Manickchand-Heileman, S., Paya, I., Sainsbury, K., Arreguin_Sanchez, F. and Zwanenburg, K. (2000) Impact of fishing on size composition and diversity of demesal fish communities. *ICES Journal of Marine Science* 57, 558-71.

Christensen, V. (1995a) A multispecies virtual population analysis incorporation information of size and age. *ICES CM* 1995/D:8.

Christensen V. (1995b) Ecosystem maturity - toward quantification, *Ecological Modelling* 77, 3-32.

Christensen, V. (1998) Fishery-induced changes in marine ecosystems: insights from the Gulf of Thailand. *Journal of Fish Biology* 53 (Suppl.A), 128-43).

Christensen, V. and Pauly, D. (1992) the ECOPATH II-a software for balancing steady-state ecosystem models and calculating network characteristics. *Ecological Modelling* 61, 169-85.

Christensen, V. and Pauly, D. (eds) (1993) Trophic models of aquatic ecosystems. ICLARM Conference Proceedings No. 26. International Center for Living Aquatic Resources Management, Manila.

Christensen, V. and Pauly, D. (1998) Changes in models of ecosystems approaching carrying capacity. *Ecological Applications* 8 (1), 104-9.

Christensen, V., Walters, C.J. and Pauly, D. (2000) *Ecopath with Ecosim: a User's Guide*, Fisheries Center, University British Columbia, Vancouver: and Penang: International Center for Living Aquatic Resources Management.

Cortes, E. (1999) Standardized diet compositions and trophic levels of sharks. *ICES Journal of Marine Science* 56, 707-17.

Cotte, M.J. (1994) *Poissons et animaux aquatiques aux temps de Pline: Commentaires sur le Livre IX de 'Histoire Naturelle de Pline*, Paris. Paul Lechevalier.

Cousins, S.H. (1995) The trophic continuum in marine ecosystems: structure and equations for a predictive model In: R.E. Ulanowicz and T. Platt (eds), Ecosystem theory for biological oceanography, *Canadian Bulletin of Fisheries and Aquatic Science* 213, 76-93.

Darwin, Charles (1859) *The Origin of Species*. London: John Murray.

Frocese, R. and Pauly, D. (2000) *Fishki Base 2000: Concepts, Design and Data Sources*. Los Banos, Philippines: ICLARM.

Gause, G.F. (1934) *The Struggle for Existence.* Baltimore, Md.: Williams and Wilkins.

Gayanilo, F.C., Jr., Sparre, P. and Pauly, D. (1996) *The FAO-ICLARM Stock Assessment Tools (FiSAT) User's Guids.* FAO Computerized Information Series/Fisheries **8**.

Golley, F.B. (1993) *A History of the Ecosystem Concept in Ecology: More than the Sum of its Parts.* New Haven, Ct.: Yale University Press.

Grigg, R.W. (1982) Darwin Point: a threshold for atoll formation, *Coral Reefs* 1, 29-34.

Ivlev, V.S. (1961) Experimental ecology of the feeding of fishes. New Haven, Ct.: Yale University Press.

Jones, R. (1982) Ecosystems, food chains and fish yields. In.: D. Pauly and G.I. Murphy (eds) *Theory and Management of Tropical Fisheries.* ICLARM *Conference Proceedings, 9, pp.* 195-239. International Center for Living Aquatic Resources Management, Manila.

Kirkwood. G.P. (1982) Simple models for multispecies fisheries, In: D. Pauly and G.I. Murphy (eds) *Theory and Management of Tropical Fisheries.* ICLARM Conference Proceedings 9, pp. 83-98. International Center for Living Aquatic Resources Management, Manila.

Kline, T.C. Jr. and Pauly, D. (1998) Cross-validation of trophic level estimates from a mass-balance model of Prince Sound using 15N/14N data. pp. 693-702 In: T.J.Quinn II, F. Funk, J. Heifetz, J.N. Ianelli, J.E. Powers, J.F. Schweigert, P.J.Sullivan and C.-I. Zhang (edgs). *Proceedings of the International Symposium on fishery Stock Assessment Models.* Alaska Sea Grant College Program Report No. 98-01. Alaska Sea Grant, Fairbanks.

Koslow, J.A. (1997) Seamounts and the ecology of deep sea fisheries. *American Scientists* 85 (2), 168-76.

Laevastu, T. and Larkins, H. (1981) *Marine Fisheries Ecosystem: Its Quantitative Evaluation and Management.* Farnham, Surrey: Fishing News Books.

Larkin, P.A. and Gazey, W. (1982) Application of ecological simulation models to management of tropical multispecies fisheries. In: D. Pauly and G.I. Murphy (eds) *Theory and Management of Tropical Fisheries.* ICLARM Conference proceedings 9, pp. 123-40. International Center for Living Aquatic Resources Management, Manila.

Lindeman, R.L. (1942) The trophic dynamic concept in ecology. *Ecology,* 23 (4), 399-418.

Lotka, A.J. (1925) *Elements of Mathematical Biology.* New York: Dover Publications.

MacDonald, J.S. and Green, P.H.(1983) Redundancy of variables used to describe importance of prey species in fish diet. *Canadian Journal of Fisheries and Aquatic Sciences* 40, 635-7.

MacKay, A. (1981) The generalized inverse. *Practical Computing* (September), 108-10.

Mathisen O.A. and Sands, N.J. (1999) Ecosystem modeling of Becharof Lake, a Sockeye salmon nursery lake in Southwestern Alaska. pp. 685-703 In: *Ecosystem Approaches for Fisheries Management.* Alaska Sea Grant College Program AK-SG-99-01.

Minagawa, M. and Wada, E. (1984) Stepwise enrichment of ^{15}N along food chains: further evidence and the relation between, ^{15}N and animal age. *Geochimica and Cosmochimic Acta* 48, 1135-40.

Munro, J.L (1983) Assessment of the potential productivity of Jamaican waters. In: J.L. Munro (ed). *Caribbean Coral Reef Fisheries Resources.* ICLARM Studies and Reviews. 7, pp. 223-48. International Center for Living Aquatic Resources Management, Manila.

Murawski S.A. Lange, A.M. and Idoine, J.S. (1991) An analysis of technological interactions among Gulf of Maine mixed-species fisheries. pp. 237-52. In: Daan and M.P. Sissenwine (eds) *Multispecies Models Relevant to Management of Living Resources.* ICES Marine Science Symposia (193), pp. 237-52.

NRC (1999) *Sustaining Marine Fisheries.* National Research Council. Washington DC: National Academy Press.

Odum, E.P (1969) The strategy of ecosystem development. *Science 164,* 262-70.

Odum, W.E. and Heald, E.J. (1975) The detritus-based food web of an estuarine mangrove community, In: L.E. Cronin (ed), *Estruarine Research.* Vol. 1 New York: Academic Press, pp. 264–86.

Opitz, S. (1993) A quantitative model of the trophic interactions in a Caribbean coral reef ecosystem. In: V. Christensen and D. Pauly (eds) *Trophic Models of Aquatic Ecosystems.* ICLARM Conference Proceedings. 26, pp. 259-67.

Pauly, D. and Christensen, V. (1995) Primary production required to sustain global fisheries. *Nature* (317), 155-7.

Pauly, D, Soriano-Bartz, M.L. and Palomares, M.L.D. (1993) Improved construction, parametrization and interpretation of steady-state ecosystem models. In: V. Christensen and D. Pauly (eds) *Trophic Models of Aquatic Ecosystems.* ICLARM Conference Proceeding, 26, pp. 1-13.

Pauly, D., Trites, A., Capuli, E. and Charistensen, V. (1998a) Diet composition and trophic levels of marine mammals. *ICES Journal of Marine Science* 55, 467-81.

Pauly, D., Froese, R. and Christensen. V. (1998) How pervasive is fishing down marine food webs': response. *Science* 282 (5393), 1383a, 20 November.

Pauly, D., Christensen V., Dalsgaard, J., Forese, R. and Torres Jr, F.C. (1998c) Fishing down marine food webs. *Science* 279, 860-3.

Pauly, D., Christensen, V. and Walters, C. (2000) Ecopath, Ecosim and Ecospace as tools for evaluating ecosystem impact of fisheries. *ICES Journal of Marine Science* 57, 697-706.

Pimm, S.L. (1982) *Food Webs,* Population and Community Series. London: Chapman & Hall.

Polovina, J.J. (1984) Model of coral reef ecosystem I. The ECOPATH model and its application to French Frigate Shoals. *Coral Reefs.* 3, 1-11.

Polovina, J.J. and Tagami, D.T. (1980) Preliminary results from ecosystem modeling at French Frigate Shoals. In: R.W.Griggs and R.T. Pfund (edg) *Proceedings of the Symposium on the Status of Resources-Investigations in the Northwestern Hawaiian Islands,* 24-25 April (1980) University of Hawaii, Sea Grant Misc. Reports UNIHI-SEAGRANT-MR-80-04, pp. 286-98.

Pope, J.G. and Knight, B.J. (1982) Multispecies approaches to fisheries management In: M.C. Mereer (ed) *Canadian Special Publication of Fisheries and Aquatic Science,* no. 59, pp. 116-18.

Rigler, F.H (1975) The concept of energy flow and nutrient between trophic levels. In: W.H. van Dobben and R.H. Lowe-McConnel (eds). *Unifying Concepts in Ecology.* The Hague: W. Junk B.V. Publishers, pp. 15-26.

Sheldon, R.W., Prakash, A. and Sutcliffe, W.H. (1972). The size distribution of particles in the ocean. *Limn. Oceanogr.* 17,327-40.

Shin, Y. (2000) Interaction trophiques et dynamiques des populations dans les ecosystems marins exploites: approache par modelisation-individus-centree. Thesis, Universite Paris 7,245 pp.

Shin, Y. and Cury, P. (1999) OSMOSE: a multispecies individual-based model to explore the functional role of biodiversity in marine ecosystems. In: *Ecosystem Approaches for Fisheries Management.* Alaska Sea Grant College Program AK-SG-99-01, pp. 593-608.

Thienemann, A.F.(1925) Der See also Lebenseinheit. *Naturwissenschaften* 13,589-600.

Thompson, D.W. (translator) (1910) *Historia Animalium,* Vol. 4. In: J.A. Smith and W.D. Ross (eds) *The Work of Aristotle.* Oxford: Clarendon Press.

Tort, P.(ed) (1996) *Dictionnaire du Darwinisme et de L' Evolution,* Paris: Presses Universitaires de France.

Trites, A.W., Livingston, P.A., Mackinson, S., Vasconcellos, M.C., Springer, A.M. and Pauly, D. (1999) Ecosystem Changes in and decline of marine mammals in the Eastern Bering Sea: testing the ecosystem shift and commercial whaling hypothesis. *Fisheries Centre Research Reports,* 7(1).

Ulanowicz, R.E. (1986) *Growth and Development: Ecosystem Phenomenology.* New York: Springer Verlag.

Ursin, K. (1973) On the prey preference of cod and dab. *Meddelser fra Danmarks Fisheri og Havundersogelser* N.S. 7, 85-98.

Vasconcellos, M., Mackinson, S., Sloman, K. and Pauly, D. (1997) The stability of trophic mass-balance models of marine ecosystems: a comparative analysis, *Ecological Modelling* 100, 125-34.

Volterra, V. (1926) Variations and fluctuations of individuals of animals living together, pp. 409-48 In: R.N. Chapman (ed) (1931) *Animal Ecology. with Special Reference to Insects.* New York: Mc-Graw-Hill.

Walters, C.J. and Juanes, F. (1993) Recruitment limitation as a consequence of natural selection for use of restricted feeding habitats and predation risk taking by juvenile fishes. *Canadian Journal of Fisheries and Aquatic Sciences* 50, 2058-70.

Walters, C., Christensen, V. and Pauly, D. (1997) Structuring dynamic models of exploited ecosystems from trophic mass-balance assessments. *Reviews in Fish Biology and Fisheries* 7 (2), 139-72.

Walters, C., Pauly, D. and Christensen, V. (1998) Ecospace: prediction of mescosale spatial patterns in trophic relationships of exploited ecosystems, with emphasis on the impacts of marine protected areas. *Ecosystems.* 2, 539-54.

Walters, C.j., Pauly, D., Christensen, V. and Kitchell, J. (1999) Representing density dependent consequences of life history strategies in aquatic ecosystems: Ecosim II. *Ecosystems* 3, 70-83.

Wigner, E. (1960) The unreasonable effectiveness of mathematics in the natural sciences, Communications on Pure and Applied Mathematics, 13, 1-13.

Chapter 17

SOME MODELS INDIVIDUAL BASED

All life stages of fish population varies. This is mainly caused by spatial and temporal fluctuations in the environment, which in turn result in variation in feeding, growth history and adaptation among individuals in a population. This individual variability often has consequences for population abundance, spatical distribution and fecundity. Traditional ecological models, such as the Lotk-Volterra competition models or the Ricker model for stock recruitment do not take into account variability among individuals. Rather, these models use population abundance as a single characteristic that defines the population dynamics and it is assumed that individuals are indentical. Consequently, features such as size structure, known to be important in fish population dynamics, are left out from the model specifications.

The traditional models have been developed into structured population models where the population is divided according to age, stage or some physiological criterion. Such structured models have proven successful for many applications in ecology and fisheries science and the use of differential equations or matrix models allows analytical solutions. Even though the structured models do partition a population, it is difficult to incorporate features such as spatial detail into these models, because it is rarely reasonable to assume that all individuals of a certain state occupy the same position in space. Structured models also generally divide the population by one variable, but real individuals may differ with regard to many variables.

These features can be implemented in individual-based models (IBM), which keep track of each individual in a population. In these models individuals can be characterized by state variables such as weight, age and length, and they also allow behavioural strategies to be implemented in a spatial context. This allows the properties of a population to be described by the properties of its constituent individuals. Model validations against data can be done at the individual level, which is an appealing property because observations often are performed on single individuals. Also, models based on individuals benefit from having the same basic unit as natural selection. These issues make individual-based modelling an appealing tool in ecology. The approach is not new per se, but rather it is analogous to the old reductionism that has been very successful in empirical sciences.

On modelling growth and survival of large mouth bass (*Micropterus salmoides*) and winter flounder (*Pseudopleuronectes americanus*) respectively, set the scene for extensive later use of IBMs in early life history studies. The major motivation for individual based modelling of these early life stages has been to explore causes of recruitment variability to commercial fish stocks. In order to simulate the survival and spatial distribution of early life stages of fish cohorts, it is important to take account of individual variability, since the eventual survivors tend to differ from average individuals at earlier stages. Studies of early life history in fish have consequently been one of the topics where IBMs have been applied most extensively. Although the individual-based modelling approach was initiated in the late 1970s, it is only since the influential review of Huston *et al.* (1988) that it has been applied extensively in ecology.

Topics that modellers of fish populations might need to deal with and then provide some relevant examples for the particular fields. As a result of this we will not cover many replicate studies dealing with similar issues, but instead present a wider range of applications taking advantage of the individual-based approach. We start out with a presentation of the IBM concept, including development and evaluation, and provide some recipes for making different kinds of IBMs. Then we move on to a review existing literature on IBMs.

SPECIFYING INDIVIDUALS IN IBMs

The Attribute Vector

Here we refer to IBMs as models that treat individuals as explicit entities, the so-called *i*-state configuration models. We will focus mostly on these models, but we will also discuss structured models that sometimes have been classified as IBMs. The *i*-state refers to individual features such as body weight, energy reserves and sex, while corresponding *p*-states represent the whole population such as population abundance and average *i*-states of population. It can be fruitful to illustrate the concept of IBMs by using an attribute vector A_i (Chambers 1993), which contains all the states αm_i used to specify an individual *i* such as age, weight, sex, hormone levels and spatial coordinates (x_i, y_i, z_i) at time *t*:

$$A_i = (\alpha 1, \alpha 2, \alpha 3, \ldots \alpha m_i, x_i, y_i, z_i, t) \tag{1}$$

The greater the attribute vector, the more differences between individuals can be specified within the model.

In structured models (Metz and Diekman 1986; Tuljapurkar and Caswell 1997), populations are divided into stages based on some key variable, for example age, which is commonly applied in, for example, the virtual population analysis of quantitative fisheries science. Using the attribute vector concept, one may describe structured models as:

$$A_j = (s_j, n_j), \tag{2}$$

where s_j is stage *j* and n_j is the number of individuals of the population in stage *j*. The changes in A_j can then be projected using models such as Leslie matrix models or partial differential equations, which can be called *i*-state distribution models (Caswell and John 1992). In IBMs, each individual is specified independently, which means that the number term n_j of equation is 1, and essentially removed from the attribute vector. However, even though the individual-based structure is appealing, it is virtually impossible to simulate even small fish stocks on a truly individual basis because of the great abundances involved. To allow the advantages of the individual-based approach and still be able to simulate large populations such as fish stocks, the super-individual approach was introduced. A super-individual represents many identical individuals and in this case the number of such identical siblings (n_s) thus becomes an attribute of the super-individual:

$$A = (\alpha 1_s, \alpha 2_s, \alpha 3_s, \ldots \alpha m_s, x_s, y_s, n_s, t_s) \tag{3}$$

where A_s is the attribute vector of super-individual *s*. Mortality operates on the super-individual and the number of siblings of each super individual is thus decreased in proportion to the mortality rate. This is an efficient way of maintaining the individual-based structure, and still be able to simulate the large population sizes that occur in natural populations. When the n_s gets below a threshold value, the way the mortality rate operates can be changed to probabilistic mortality using Monte Carlo techniques for the remaining siblings. Populations with high mortality, such as fish populations, can effectively be simulated by replacing a dead super-individual through random resampling from the live portion of the population. The internal number n_s of the donor individual is then divided in two and the dead super-individual inherits the attributes of the donor. Thus one may keep the number of super-individuals constant while changing the number of individuals that actually are represented by each super individual.

The aggregation in super-individuals has some obvious similarities with the structured models, Tuljapurkar and Caswell 1997. However, the structured population models are based on partial differential equations or Leslie matrices, while the superindividual approach still maintains the same structure and representation of processes as in the IBMs discussed above. In structured models all stages are usually assumed to experience the same environment the therefore to respond similarly (Caswell 1996). The super-individual approach is more flexible than and structured models and allows a simple way of scaling IBMs to realistic fish stock abundances. Structured models and configuration models may differ in their applicability depending on the topic in question, but in cases where both approaches can be applied they tend to give similar predictions.

The Strategy Vector

Real individuals have adaptive traits, such as life history and behavioural strategies that specify how they should live their life. The previous lack of IBM studies involving life-history strategies and behaviour of individuals could due in part to a lack of appropriate techniques for implementing these features. However, adaptive traits can be modelled by introducing a strategy vector, S_i that specifies the adaptive traits, such as life-history traits or behaviour, of an individual:

$$S_i = (\beta 1_i, \beta b2_i, \beta 3_i, \ldots .\beta m_i), \tag{4}$$

where βm_i is the adaptive trait m of individual i. The strategy vector may be considered as analogous to a biological chromosome as in the genetic algorithm (Holland 1975), but, S_i may also be updated during the individual's life as a way to simulate learning. For example, this can be done using reinforcement learning by allowing rewards for unprofitable ones. This process allows the individual to produce increasingly more favourable behaviours as it learns about its environment. The combination of attribute vectors and strategy vectors thus enables most relevant characteristics of individuals to be implemented in IBMs.

Features of Individual-Based Models

IBMs can be classified by the degree to which different factors such as book-keeping (*i.e.* the continuous update of the attribute vector), space, adaptive traits and local interactions are specified in the model (Fig. 1). Depending on the nature and scale of the problem of interest one may develop a simple book-keeping model, of complex models taking into account more aspects of individuals. All IBMs contain a book-keeping procedure, but they need not contain the other features listed in Fig. 1. Another common feature of IBMs is to represent ecological processes by mechanistic models.

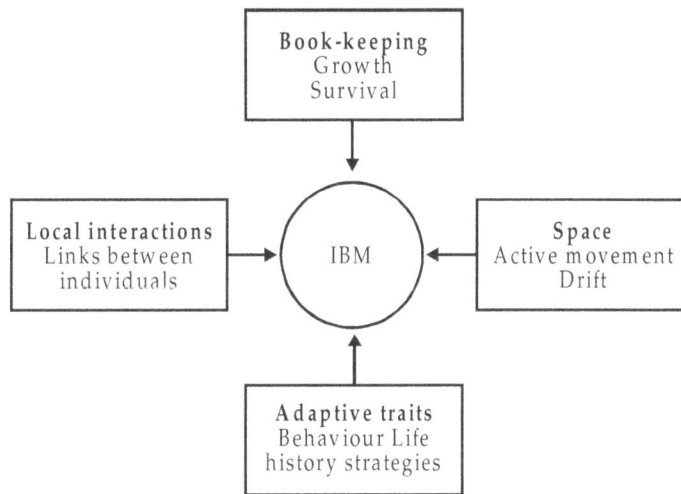

Fig. 1 : Schematic classification of various types of IBMs. Most IBMs keep track of growth and survival of individuals in a population. Implementation of space allows a wider range of studies to be undertaken. A further increase in the complexity and flexibility of models to incorporate adaptive traits such as behaviour and life history strategies may further promote our understanding of how individuals relate to each other. The simulation of local interactions involves a high resolution both in spatial detail and behavioural actions.

Mechanistic models

As opposed to empirically fitted models, mechanistic models aim at representing the actual process that is taking place in more detail. Thus instead of simply fitting growth rate as an empirical relationship of, for example, size, a mechanistic model of growth will address the various processes involved. These include encounters with prey the ingestion process and bioenergetic. Mechanistic models are used extensively in IBMs to represent individual processes such as perception, predator and prey encounters, and bioenergetics. These are indicated by bold squares in Fig. 2, which shows a flow chart for events related to feeding, growth and predation of fish larvel. In addition to these individual processes, the environment is also often represented through mechanistic models of features such as light and ocean circulation. For a thorough discussion of mechanistic models in fish biology.

Book-keeping

As shown above, this is facilitated using the attribute vector. The entire population is tracked using an attribute matrix with dimensions equal to the number of attributes times the number of individuals in the population. All events that may occur in a period of a time step will be addressed sequentially in the IBM, as if they appeared one by one, and it may be important to analyse the processes in a particular sequence. Organism may be eaten after it has starved to death, it may not starve after having been eaten. The conceptual framework for bookkeeping illustrated in Fig. 2 is characteristic of most IBMs, which deal with growth and survival of fish. Monte Carlo simulations are continuously used to decide the outcome of feeding and mortality processes.

Monte Carlo Simulations

Monte Carlo simulations are often applied in IBMs, but this approach is rarely well defined. In general, Monte Carlo techniques involve drawing random numbers from some probability distribution, in a way similar to gambling situations from which the name derives. The usual way to apply this concept in IBMs is to assume a probability for some event and then draw a number from a random-number generator to determine the outcome. For example, one might a certain probability that an individual dies during a time step. A random number is then drawn, and if the random number is smaller than the mortality risk, the individual dies. Given that the random numbers are uniformly distributed, the probability of drawing a number within an interval is equal to size of the interval. If the mortality risk is 0.1, there is a 10% probability that the random number will be between 0 and 0.1, and a similar probability of dying. This can be carried out to determine whether an individual survives, eats food or obtains mating and so forth.

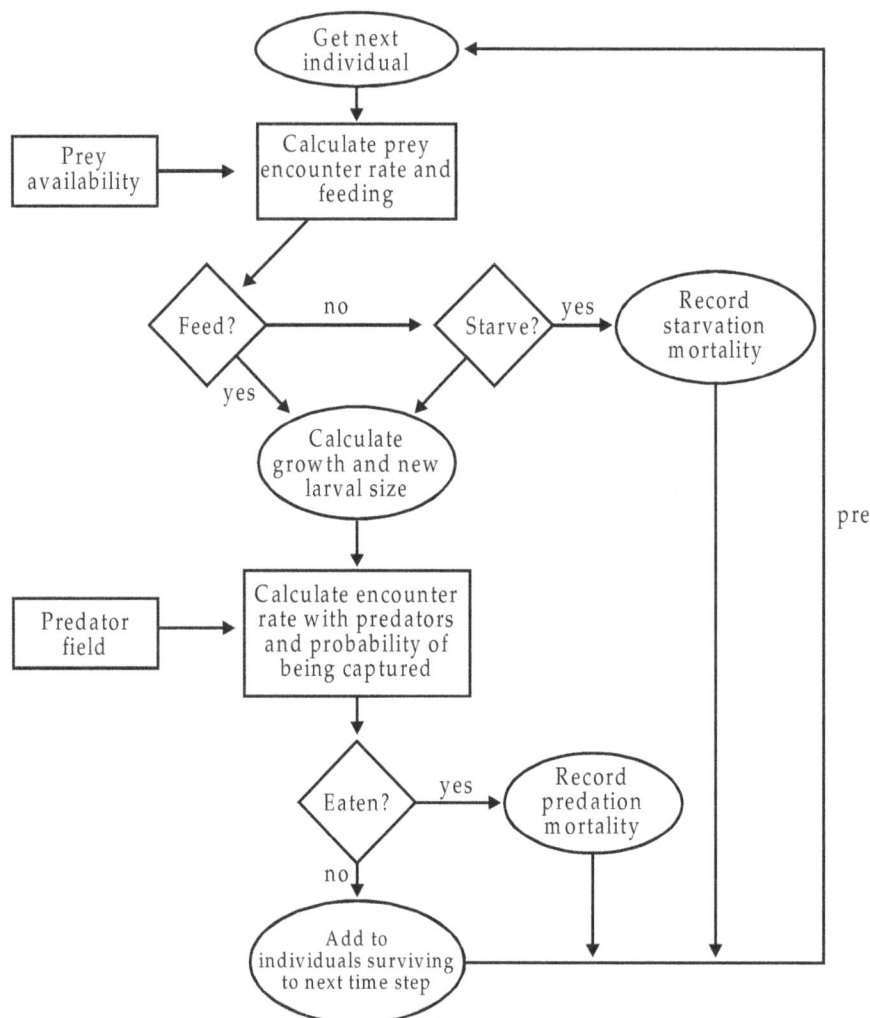

Fig. 2 : A conceptual model of a sequential representation of events related to growth and survival of fish larvae. The same sequence of events must be repeated for every individual in the population, before starting again with the first individual in the next time step. Bold squares indicate where mechanistic models are applied and ovals indicate where book-keeping is performed. (Modified from Crowder *et. al.* 1992).

Spatial Detail

Traditional ecological models specify space as a homogenous well-mixed compartment (Fig. 3). In metapopulation models, space is implemented as habitable patches surrounded by non-habitable areas. Metapopulation models have been applied to some problems in fisheries science, for example to explain population structuring in herring stocks. A still finer spatial resolution is to arrange space in two- or three-dimensional grids where essentially any spatial scale can be represented (Fig. 3). IBMs can be used in all these spatial arrangements.

The development of IBMs has made it possible to implement space in ecological models in more realistic manner than structured models allow. Grid models have been extensively applied in fisheries science for studies of the distribution of eggs and larvae. An example of how such model are set up is provided below.

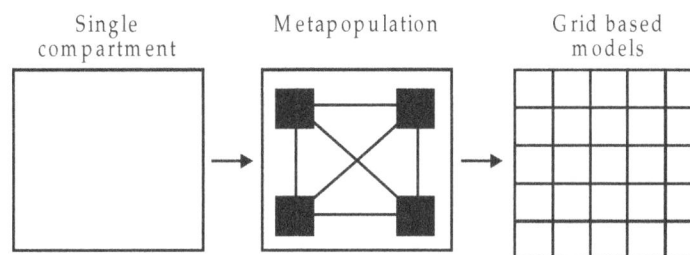

Single compartment   Metapopulation   Grid based models

Fig. 3 : Different ways of representing spatial detail in models.
The complexity in spatial representation increases in the direction of the arrows.

Recipe for Simulating Drift of Larval Fish

The flow chart in Fig. 2 does not contain any decisions made by the organism. This situation may be relevant for the planktonic larval stages. Spatial positioning of fish during this life stage is determined by their buoyancy, the turbulence of the environment and the advective transport of water. In lakes, however, the impact of advective forces is much smaller than in the sea and in rivers, where larval drift is an integrated part of the life cycle of many fishes. Although assuming a spatially homogenous compartment may be a reasonable simplification for some lakes, it is generally not so for ocean systems.

Models for marine fishes have therefore increasingly been developed to incorporate advective fields and temperature data from ocean circulation models. These models are grid based with a resolution typically of 1-20 km. Physical ocean circulation models provide current velocity vectors for each grid cell at each time step. These vectors are used to move individuals (or 'particles', representing drifting biological organisms) about (*e.g.*). The drift path may also be impacted by swimming of the larvae, which is then added to the physical vector. Once the drift path of a larva is found, the temperature, food, light intensity etc. along this path can also be estimated (Table 1).

Table 1 : Pseudo-code for implementing spatial detail in studies of larval fish.

1. Create current vector from a hydrodynamic model.
2. Add component of active individual movement.
3. Integrate velocity vector over the duration of the time step to produce individual trajectories.
4. Find temperature, food concentration, light intensity, predator density, etc., along trajectory.
5. Estimate food encounter rate, predation risk, bioenergetics etc.
6. Find state (spatial position, physiological state, alive/starved/eaten) of organism at the end of time step.

Adaptive Traits

As mentioned above, the strategy vector S_i can be used to specify the adaptive traits of an individual. In some studies IBMs are simply tailored to match the life history of target species, in which case behaviour is not specified explicity in the model, but the consequences of certain behavioural strategies for growth and mortaily may still be simulated. In other cases, finding the best behavioural trait is part of the modelling exercise itself. We mainly discuss behaviour below, but life-history strategies can often be implemented in the same manner. In general three approaches prevail for specifying behaviour in IBMs: rule-based approaches, optimization, and adaptation.

Rule-based Approaches

Even though rules may be chosen based on their evolutionary, profitability, the essential part of the rule-based approach is that the rules are provided by the modeller and that behaviour results directly from the specified rule. One example of a behavioural rule is random walk, which simply is to move individuals about in a tandom fashion with equal probability of going in each direction at each time step. Such a simple behavioural procedure resembles a natural situation in many cases of local searching in animals. Other rule-based concepts specified in models are the use to taxis and kinesis, which specify reactions to stimulti be orientating relative to stimuli position or responding in proportion to stimuli intensity respectively.

Optimization

The second approach to modelling behaviour is generally referred to as the 'optimization approach Parker and Maynad Smith 1990. This approach underlies many of the discussions of fish behaviour in Volume 1, as exemplified by Mittelbach's discussions of foraging theory in Chapter 11, Volume 1. The optimization approach in behavioural ecology is ultimate, based directly on the survival value of behavioural traits. Stochastic dynanmic programming (SDP) Mangel and Clark 1988, Houston and McNamara 1999; Clark and Mangel 2000. where the entire solution space is sought and the best solution is chosen, is an example of the optimization approach. SDP relates to states rather than individuals and the optimal strategy is calculated for state (for example, weight or energy level) at each time step in a backward iteration. This allows state variability and state-specific behaviour to emerge. Once the backward iteration procedure is complete, an IBM can be used to simulate individual trajectories form the start to the horizon of the model. SDP models may use Monte Carlo simulations for determining growth and reproduction, which makes this approach individual-based (Clark and Mangel 2000). Traditionally, population growth rates such as the instantaneous rate of increase r or the lifetime reproductive success R_0 have been applied as fitness measures in optimization models (Roff 1992; Hutchings, Chapter 7, Volume 1). These measures are well suited for situations where the entire lifesepan of animal is considered. However, it is more difficult to assess type profitability of behaviours at an instant. Solutions of this problems can be achieved using 'rules of thumb' that the generated as predictions from evolutionary ecology, such as the ideal free distribution, the marginal value theorem and Gilliam's rule. These concepts are applied to forging and to habitat choice by Mittelbach. Given the assumptions, these approaches will provide optimal solutions to the respective behavioural problems.

Adaptation

Adaptive models find 'good' behavioural strategies by using gradual improvements in behaviour through simulated evolution or learning. When simulating evolution, the behavioural strategies are coded numerically on the strategy vector (equation 4) and passed on from parents to offspring. This technique is known as the genetic algorithm (GA) Holland 1975). In the GA a specific measure of fitness such as R_0 can be used to determine which strategy victory and hence which individuals are the best, and these become parents for the new generation. Variability in the new strategy vectors are provided through recombinations among parents, and mutations. As mentioned above, reinforcement learning can be used to update behaviours within the lifetime of the individual. This approach can also be used in combination with evolved strategies. When the environment changes markedly on a short time-scale, for example within generations, it can be profitable to allow adaptation through learning. In an artificial neural network (ANN) model this would mean that the weights are changed within the life of individuals, and not only between different generations. Artificial neural networks (ANNs) apply neurobiological principles of synaptic brain activity to perform systematic output by differential weighting of input variables.

Table 2 : Recipe for incorporating behaviour and spatial detail in an IBMs of a planktivorous fish. The pseudo-code illustrates how the model is executed in a sequential manner, using a programming language such as FORTRAN.

---

1. Initiate attribute and strategy vectors for each individual
2. Year loop
   Create physical environment for the entire year light, temperature, currents
   Initiate food and predator distribution.
3. Day loop
   Update distribution and abundance of food, competitors and predators
4. Individual loop
   Perform behavioral actions
   Determine growth and mortality
   Update abundance of food items after feeding.
   Reproduce if criteria are fulfilled.

An alternative approach for evolving behaviour in models is to use so-called emergent (endogenous) fitness, which means that individuals with strategy vectors live and reproduce in an evolving population (Strand *et al.* in press). Thus rather than maximizing a specific fitness measure, those individuals who manage to pass on their own strategies by reproducing with other individuals in the population will be the most fit. However, the degree to which this reflects a natural system depends upon the way the environment is specified in the particular model system.

Recipe for Adaptive Models with a Strategy Vector

IBMs containing both spatial detail and behaviour are generally very complex, since they often involve a range of mechanistic models for specifying temperature, drift, feeding and behaviour. The following recipe describes the conceptual flow of a model containing both attribute and strategy vectors and spatial detail. It is based on the model of Huse (1998), where fish behaviour and life history strategies for a planktivorous fish are adapted over many consecutive generations (Table 2).

1. The attribute and strategy vectors are initialized for each individual in the population. While the attribute vectors are set to 'common' values, the strategy vectors are initiated randomly within certain intervals. The attribute vector is specified as A_i (age, weight, energy level, position), and the corresponding strategy vector is: S_i (timing of spawning location, energy allocation rule, size at maturity, movement). Movement behaviour is

determined using an (ANN and its weights are implemented in the strategy vector.

2. At the start of the year the environment for the coming year is established. As in the previous recipe, temperature and current fields are produced by an ocean circulation model. The predator abundance is assumed to increase linearly with increasing temperature. The food distribution is initiated and is consistently updated below.

3. The day loop runs over each day of the year, and, most importantly, the food distribution is updated according to the production and import from advective transport provided by the physical model.

4. In the individual loop most of the biological features of the IBM are implemented. The first task is to infer morality, since the model uses the super-individual approach, a proportion of the number of clones is removed according to the predation risk at the present location. In the same fashion one may include removal of fish caused by fishing mortality. Food intake is determined from local encounters with food, and growth is then calculated using the bioenergetic model of Hewett and Johnson (1992). For adults, surplus energy is divided among growth and reproduction according to the individual allocation strategy. Juveniles, on the other hand, are assumed to put all their energy into growth. At an individually specified spawning time and location, individuals may reproduce given that certain criteria are fulfilled. The new individuals inherit behavioural and life history strategies from their parents by recombination with a probability for mutations, to mimic biological reproduction. These are essential parts of the genetic algorithm. At the end of the individual loop, movement is determined using the ocean drift for the planktonic larvae (see recipe above) or the ANN for fish above a certain size. The ANN calculates movement from information about the local abundance of predators, growth, temperature and position.

These points make up the basic structure of a spatial life-history model of fish. The model is run for a large number of years (300-500) and generates life-history and migration strategies that resemble observed migration patterns. Although the example is for capelin migrations in the Barents Sea, the model approach is general and can be applied to virtually any fish stock. It has also been applied successfully for simulating vertical migration in mesopelagic fish.

What is the best modelling approach?

What type of model is 'best' for implementing adaptive traits, or which one of these techniques should one use? The answer is, as often -'it depends'. Some general recommendations emerge from the discussion above. In terms of specification of adaptive traits in IBMs, the strategy vector and the adaptation concept is appealing in many ways. The advantage of this concept is its generality because it can encompass most ecological processes including density dependence, state dependence and stochastic environments. Furthermore, when using ANNs it is possible to simulate behaviour form stimuli, thereby allowing the use of conventional behavioural terminology and perspective. The downside of the adaptive is that it is impossible to know whether the optimal solution is found unless this is calculated by other means. This is, however, ensured using optimally models such as SDP, which is one of the great advantages of that approach. Another advantage of SDP is the ability of this technique to include individual state and time constraints in the optimization criterion. A conclusion can therefore be that if the biological question involves state dependence or making sure that the optimal strategy is found, then SDP should be used. On the other hand, if the study involves problems that have high dimensionality and/or include stochastically or density dependencies, then the adaptive models will be the best approach. If one is interested in how large-scale patterns and/or complex phenomena emerge from individual behaviour, it can be productive to apply simple behavioural rules.

Local interactions

In some cases, for example in fish schooling (Reynolds 1987; Vebo and Nottestad 1997; Stocker 1999), behaviour is dependent mainly upon what conspecifics and predators in the vicinity are doing. A simple modelling approach that takes local interactions into account is cellular automata. Under this approach, the modeller defines strict rules that are similar for each individual cell in a lattice. The rules then specify how the automata change state according to the state of their surrounding cells. The emerging pattern of the lattice then results from the local interactions among the automata.

FORMULATING AND TESTING IBMS

Model Formulation

Since IBMs typically are built from an extensive set of submodels, there are many things that can go wrong model formulation. It is therefore important to develop a common framework for putting together IBMs. Railsback (2001) provides six points that should be considered when formulating an IBM: (1) Emergence: what processes should be imposed by empirical relations and what should emerge from mechanistic representations? (2) Adaptive traits: what kind of adaptive processes should be included in the model? This point has to be related to the spatial scale and the major questions being addressed. (3) Fitness measure: what is the appropriate fitness measure for the adaptive traits of the model? (4) State-based dynamics: how should decision depend on individual state? (5) Prediction: what are realistic assumptions about how animals predict the consequences of decisions? (6) Computer implementation: what user interface is necessary for implementing, validating and testing IBMs? In addition to these points it is important to provide results that can be tested against observations. Although testing against real data is not always necessary, since models may provide valuable insight through sensitivity analyses, it is most often an advantage in model development. The points shown above are important to keep in mind when constructing an IBM, and we shall return to some of these as we go along. Other things to keep in mind when formulating a model are that the choice of complexity and model structure should be based on the level of understanding of the environmental and biological processes operating. Lastly a model is at best a highly simplified but biased representation of nature, and uncertainty can be reduced by attacking the problem with several models what differ in assumptions and structure.

Evaluation of IBMs

The evaluation process can be divided into verification and validation. Verification is the process of checking that the internal logic of the computer model is correct and that the model actually does what it is intended to do. This process is performed continuously as a model is developed. Validation, on the other hand, aims at determining the ability of the model in describing observed phenomena. This process can be divided further into checking the validity of parameter values, and secondary and primary model predictions. The validity of parameter values is traditionally tested through a sensitivity analysis. Sensitivity analyses involve varying parameter values and studying the effect on model output. For IBMs, which usually have a great number of parameters, it is simply not feasible to test the sensitivity of all these. Rather some 'key' parameters should be chosen for testing. Since IBMs tend to be composed to submodels. Many features of an IBM can be tested. This is what is referred to as secondary model predictions. With regard to a fish model this can be, for example, testing the performance of a bioenergetic model used, even though the primarily aim of the model is to provide population dynamics of a target species. The primary and secondary model predictions can then be tested independently of each other. A consequence of model evaluation is either to accept the model's performance or, alternatively, to try to revise parts of the model that produce erroneous predictions. One should, however, be careful not to make the model fit observations by changing parameter values or submodels uncritically. This is especially important for IBMs, which typically simulate processes in a mechanistic manner using a large number of parameter values.

When it comes to software implementation, any kind of programming language can be used, but object-orientated languages such as C⁺⁺ are especially well suited for constructing IBMs. In addition to providing a nice structured programming with individuals as the basic units, these languages generally also provide good visualization opportunities. For example, the object-orientated Swarm package is tailored for individual-based simulations and provides a number of features for visualization of results and book-keeping of individuals and processes. Using Swarm, several aspects of the model may be monitored during development and evaluation. Swarm, which is a shareware product, is currently available for the Objective C and Java languages. Another software package especially developed for individual-based simulations is ECOSIM (Lorek and Sonnenschein 1998).

Review of Individual-Based Models in Fisheries Biology

Table 3 : Features characterizing different aspects of the early life history of fish.

What characterizes...	Variables	References
The survivors?	egg quality	Kjesbu *et al.* (1991)
	birth date	Schultz (1993)
	birth position	Berntsen *et al.* (1994); Slotte and Fiksen (2000)
	egg size	Kuntsen and Tilseth (1985)
	development	Blaxter (1986)
	prey encounter	Fiksen and Folkvord (1999)
	bioenergetics	Crowder *et al.* (1992); Fiksen and Folkvord (1999)
the environment?	food concentration	Cushing (1990); (1996)
	small-scale turbulence	Sundby and Fossum (1990); MacKenzie *et al.* (1994)
	water	Bartsch *et al.* (1989); Berntsen *et al.* (1994);
	transportation	Hermann *et al.* (1996)
	ocean climate	Cushing (1996); Anderson and Piatt (1999)
	light	Miner and Stein (1993); Fiksen *et al.* (1998)
	turbidity	Chesney (1989); Fortier *et al.* (1996)
	temperatue	Houde (1989); (1997)
	predators	McGurk (1986); Bailey and Houde (1989); Cowan *et al.* (1996)

The ontogenetic development in fishes from eggs to maturity typically involves manyfold increases in body size associated with discrete changes in morphology and increased behavioural repertoires. The way individuals vary therefore changes through the life cycle. This suggests that modelling should focus on different aspects of life at different times of the life cycle, and that each ontogenetic stage might be specified differently. For example, in the period of change from intrinsic to extrinsic energy uptake, each prey caught by a larval fish is very important, which makes food gathering a vital process to simulate in models. During overwintering or spawning, on the other hand, feeding will be a virtually unimportant activity. In the following we will review studies using IBMs in fish biology by using the structure of Fig. 4. Hence we initially discuss relatively simple models, and then add features to the IBMs as we go along. The focuses on growth and survival in early life-history but greater relevance to older life stages.

Models of Growth and Survival

For most, if not all fish species, mortality at the egg early larval stage is enormous compared to later in life. Therefore, as early as 1914, Hjort considered recruitment variation as a major research area in fisheries science. Recruitment variation may be caused by factors related to the state of the organism, its evironment or its parents (Table 3). IBMs allow a mechanistic representation of ecological processes, and consequently such models have been applied extensively in studying growth and survival of young-of-the-year fish. Beyer and Laurence (1980) recognized the importance of initial chance events in feeding of larval winter flounder. Since prey ingestion can be described as a Poisson process, some individuals will initially have success and capture prey while others have poor fate. Given that the ability to obtain food is size dependent there will be a positive feedback so that initially successful individuals will grow faster than the others and thus have a higher survival probability. Such scenarios are well represented in IBMs.

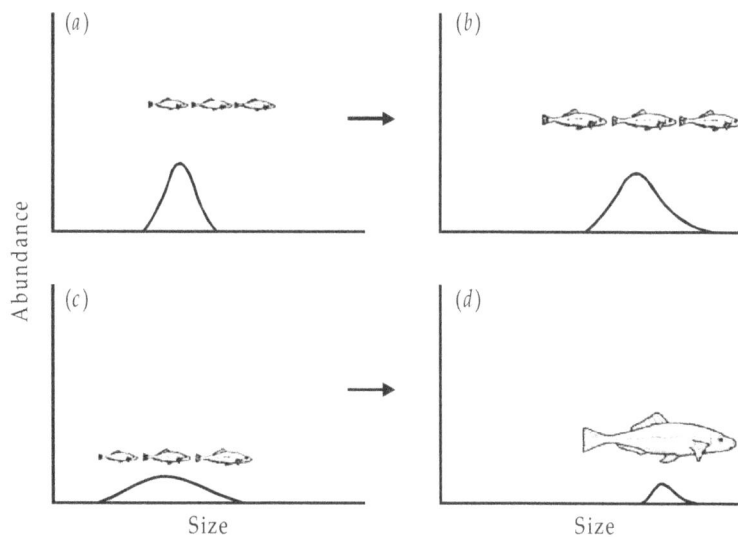

Fig. 4 : The effect of initial variation in size distribution on the outcome of cannibalistic interactions. Low initial variance (*a*) produces a cohort with homogeneous individuals at a later stage (*b*) while with a high initial variance (*c*) the biggest individuals get cannibalistic (*d*). (From Hustom *et al.* 1998; reproduced by permission of Bio Science.)

The important survival factors for eggs and larvae are prey and predator encounters, bioenergetics and organisms development. With so many factors interacting, modelling is essential in understanding how this affects individuals and populations. And since egg and larval mortality are high, models of survival should be able to focus on the lucky or clever few. The important decision variables are mainly governed by parent behaviour through choice of spawning site and time, anatomical properties such an egg size and energy density, and by developmental 'programme' displayed through the sequence of organ development. A range of ecological conditions will bring stochastically to all these variables (Table 3). The reliability of predictions from complex recruitment model will depend on how well these processes are represented mechanistically. In a classic study, De Angelis *et al.* (1979) modelled the growth of young-of-the-year large-mouth bass. They found that the initial variance in the chart was important for its development and for whether cannibalism was possible or not. In the case of low initial variance, shown by an even size distribution (Fig. 4*a*), the fish remained homogeneous with none growing big enough to eat the others (Fig. 4*b*). With large initial variance (Fig. 4*c*), however, the individuals that had largest initial size were able to become cannibalistic and eat the smaller fish, which the model run predicted to lead to survival of few but large individuals (Fig. 4*d*). It was demonstrated that the individual variability is clearly important for cohort development even though the initial average weight was the same in the two cases. The model presented by Fiksen and Folkvord (1999) provides a state-of-the-art mechanistic description of the feeding process taking into account environmental features such as small-scale turbulence, light, turbidity, temperature, prey and size structure. Alongside the model development there has been experimental work used for generating parameter values and testing of model predictions. The great detail of the mechanistic description enables this model to make more realistic predictions than model with simpler environmental description.

Models with Adaptive Traits

The simplest behavioural models are those without spatial detail. There have been some studies addressing how spawning strategies are affected by seasonal variation in temperature, growth and predation risk. Trebitz (1991) used an IBM to find the best timing and temperature for spawning with and without density dependence. Initially the spawing temperature was given randomly, but subsequent spawning strategies (temperature) were given in proportion to the survival of the strategies (individuals) until age 1. The biomass of each temperature strategy at age 1 was therefore used as a measure of strategy performance. Denisty-independent and denisty-dependent model runs predicted different optimal spawning temperatures, with a broad peak at intermediate temperatures for the density-independent situation and an almost temperature-independent profitability for the density-dependent case. This model is in many ways similar to the genetic algorithm.

Simple Spatial Systems

Natural systems are very complex. It may therefore be profitable to conduct simulation experiment, which is a computer analogy to laboratory or field experiments, in which focus can be put on a specific research topic without necessarily defining a particular natural system. A simulation experiment is especially fruitful early in theoretical development or if the problem is general rather than linked to a specific population or evironment. Tylar and Rose (1997) studies cohort consequences of different habitat choice rules in a simple spatical system life-history-based criteria such as Gillian's rule. Their results suggest that habitat choice rules have strong effects on cohort survivorship and that no single departure rule can be an evolutionarily stable strategy. They therefore concluded that the use of static life-history rules is not an appropriate way to model behaviour in a dynamic environment, as confirmed elsewhere. Tyler and Rose (1997) also found that density-dependent effects on juvenile survival can be much greater in spatially explicit models with fitness-based habitat choice than in spatially homogeneous models. SeaLab (LePage) and Cury 1997) is a spatial simulator that can be used to test hypotheses about fish reproduction and space use. This model divides space into hexagonal structures, and individuals may choose in which cell to stay and/or reproduce. LePage and Cury (1997) used SeaLab to simulate how reproductive strategies depended on the degree of variation in variable environments. An obstinate strategy, where spawning is performed under the same conditions as the individual is born, and an opportunistic strategy where individuals spawn under novel environmental conditions, were investigated. For extreme environmental variation, only populations with both the opportunistic and obstinate strategies survived. The authors discuss the results with regard to straying in natural populations. A similar artificial environment was applied by Anneville *et al.* (1998), who explored the effect of density-dependent recruitment relationships using rule-based movement in an hexagonal lattice. Their main conclusion was that local density dependence was often not detectable at large spatial scales, which stresses the importance of being explicit about scale when analysing ecological processes.

Models of Fish Distribution

Spatially explicit IBMs incorporate spatial heterogeneity, individual variability and individual movement. Although such models are complicated and necessarily consist of many modules, they provide the potential for highly realistic simulations of fish populations. Many fishes undertake extensive horizontal migrations between feeding, spawning and over-wintering areas. Such migrations involve complex interactions between individuals and the environment that are generally not well understood. Tagging experiments have provided first-hand information about the distribution of migratory fish stocks, and lately telemetric tags have enabled real-time observations of fish movements in the sea. However, it is difficult to understand the mechanisms controlling the movement of fish just from observations, and model simulations are therefore an important part of exploring the proximate and ultimate processes involved in fish migrations. Although there is likely to be some common features among fish stocks in how they move horizontally, local adaptation will be important, and it may not be trivial to transfer knowledge about the migration of one stock to other stocks. In the case of vertical migration, however, there may be more common features due to the pronounced vertical gradients and diurnal changes in light that strongly influence visual feeding and predation risk.

Larval drift was among the first topics where spatially explicity IBMs were applied. Werner *et al.* (1996) considered trophodynamics and ocean circulation in a study of cod larvae on Georges Bank by providing drift and growth trajectories of individuals. They concluded that the region of highest retention coincided with the region of highest growth, illustrating the complementary interaction between trophodynamics and circulation processes. In a similar model of walleye pollock (*Theragra chalcogramma*), Hinckely *et al.* (1996) found that the inclusion of mechanisms that determine the depth positioning of the deriving larvae are important for determining the direction of horizontal advection.

Through a combination of simple decision rules and a library of surface currents, Walter *et al.* (1997) modelled sockeye salmon (*Onchorhynchus nerka*) movement in the Northeast Pacific Ocean. Movement was simulated based on random walk, with directed swimming at certain times of the year. Their simple rules for compass bearing predict migration patterns that challenge prevailing complex models of sockeye migration. The most striking result was that inter-annual variation in the surface current led to great changes in the distribution of the salmon given that the same movement rules apply each year. The rule-based approach used by Walter *et al.* (1997) does not take into account that individual difference in state can cause differences in behaviour. State dependent habitat choice was considered by Fiksen *et al.* (1995), who used a stochastic dynamic programming (SDP) to simulate the horizontal distribution of the Barents Sea capelin (*Mallotus villous*). The model was based on prior simulation results, which specify the physical environment of the Barents Sea, and assumed distributions of predators and prey. By using a time step of one month, the optimal habitat for each state (body weight) was calculated using the SDP equations with lifetime reproductive success as a fitness criterion. The model results compared favourable with the observed distribution of the Barens Sea capelin. In contract to Walter *et al.* (1997), Fiksen *et al.* (1995) assumed that the individual capelin actively made habitat choices to maximize their Darwinian fitness. Huse and Giske (1998) and Huse (1998) used a similar environmental description as Fiksen *et al.* (1995) to simulate movement of the

Barents Sea capelin using an adaptive model with a strategy vector (equation 4) containing the weights of an artificial neural network and several life-history traits. Similarly to Fiksen *et al.* (1995), they assumed that the capelin is adapted to the environment of the Barents Sea. However, rather than assuming optimal habitat choices, an adaptive process over many hundreds of generations was simulated using emergent fitness (see above). The model was applied to study evolution of spawning areas, and it predicted, in accordance with field observations, capelin spawning to occur along the coast of Northern Norway, ANN-based models may rely on proximate sensor information, and at the same time the weights of the ANN are adapted using ultimate forces. In this way ANN models provide a link between proximate and ultimate factors in behavioural ecology. ANNs have also been applied to study tropical tuna (*Katsuwonus pelamis* and *Thunnus albacares*) migrations between the Mozambique Channel and the Seychelles using temperature data gathered by remote sensing (Dagorn *et al.* 1997). In this case the tuna was assumed to search for areas of low temperature, which are often associated with food-rich frontal zones. The ANN was used to determine which movement action the artificial tuna should make based on information about the temperature map within the daily search radius. The fitness criterion used in the GA for training the ANN was to minimize the distance to the observed arrival point near the Seychelles. As result of the adaptive process, the simulated tuna eventually.

ECOSYSTEM MODELS

Rigler (1975) presented at a key meeting of the IBP, based on the data collected by the IBP, showing that most aquatic animals, which include all those that are not strict herbivores or detritivores, feed, simultaneously, at different trophic level, Rigler's critique was extremely influential, and its echoes are still detectable. However, the solution to issues raised by this critique had a simple answer: fractional trophic levels (TL_i Odum and Heald 1975), computed, for the animals of a given population (i), from the trophic levels of all their prey (j). Thus, we have:

$$TL_i = 1 + \sum_{j=1}^{n} TL_j . DC_{ij} \tag{1}$$

where DC_{ij} is the fraction of j in the diet of i, and n is the number of prey types. Estimates of trophic level derived from equation (1) and earlier estimates of TL_i exist for marine mammals, and for sharks and other fishes (*www. fishbase, org*).

Another widely used method to estimate fractional trophic levels is through the analysis of stable isotopes of nitrogen. This relies on the observation that the ratio of ^{15}N to ^{14}N increases by about 3.4% every time proteins are ingested by a consumer, broken down and resythesized into its own body tissues, in the first study of this type, showed that trophic levels estimated from diet compositions, that is, from Ecopath models closely correlated with trophic level estimates from stable nitrogen isotopes.

The variance of trophic level estimates can also be calculated. Given the nature of TL estimates from either stable isotope ratio or diet composition studies, this variance will reflect feeding at different trophic levels, *i.e.* omnivory by the consumer under study as well as uncertainty concerning the trophic level of its food. Thus, one can define an 'omnivory index' (OI) calculated from :

$$OI = \sum_{j=1}^{n} \left(TL_j - (TL_i - 1) \right)^2 . DC_{ij} \tag{...(2)}$$

where n is the number of groups in the system, TL_j is the trophic level of prey j, TL the mean trophic level of the prey (one less than the trophic level of the predator, see above), and DC_{ij} is the fraction of prey j in the diet of predator i, again as defined above. Rigler (1975) argued that trophic levels were only a 'concept', and that mature sciences should deal with concepts only in the absence of measurable, actual entities, which alone allow testing of quantitative hypotheses. The demonstration that trophic levels estimated from diet composition data and equation (1) closely correlate with estimates from stable isotopes of nitrogen not just only cross-validates these two methods, but also establishes that trophic levels are not concepts useful for animals to various ecological groups, but actual entities, similar to the size or metabolic rate of organisms, which can be measured by different, independent methods, and whose various features, therefore, can be elements of testable, quantitative hypotheses.

Practical Uses of Trophic Levels : Tracking Food-Web Changes

Ecopath, as shown above, has estimates of trophic level as one of its outputs, along with the standard error of these trophic levels, the square root of the omnivory index. The many estimates of trophic levels that emerge from various Ecopath applications helped confirm various generalizations by Pimm (1982) and others about the structure of food webs. Also, they allowed going beyond these generalizations. Thus, Pauly and Charistensen (1995), who had assigned trophic level to all fish and in vertebrates caught and reported in FAO global fisheries statistics, could show, using between-trophic level transfer efficiencies also estimated through Ecopath

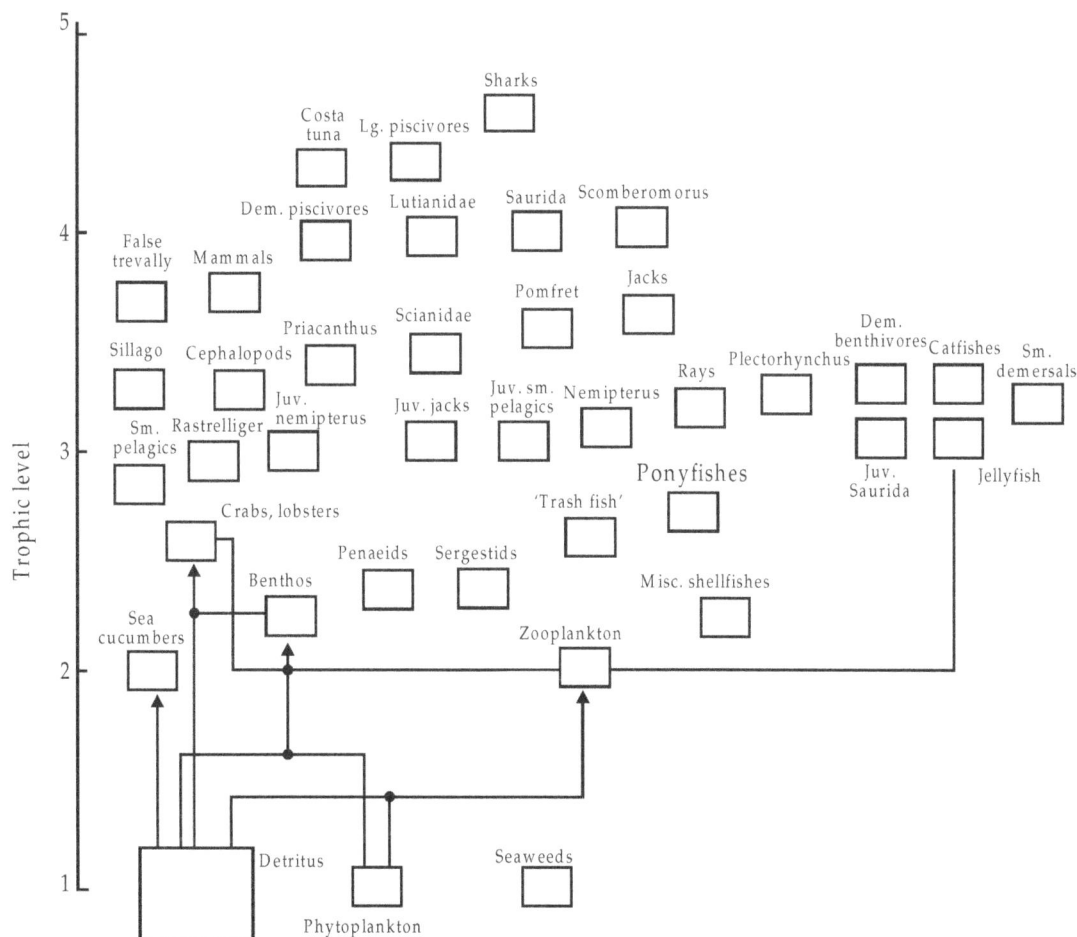

Fig. 1 : Flow diagram for a trophic model of the Gulf of Thailand. All flows between ecosystem groups are quantified, but only flows exceeding 5t-km^2 year^{-1} are indicated. The sizes of the boxes are a function of the group biomasses.

models, that the primary production required to sustain the present world fisheries was much higher than previously assumed: 8% for the global ocean and between 25% and 35% for the shelves from which 90% of the world catches originates. Also using time series of the same fisheries statistics, and trophic level for the major species, Pauly *et al.* (1998c) demonstrated a steady reduction of the mean trophic level of fisheries landings from 1950 to the present, suggesting that the fisheries increasingly concentrate on the more abundant, small, fast-growing prey fishes and invertebrates near the bottom of aquatic food webs. Both of these sets of findings, now validated through more detailed local studies quantifying human impacts on marine ecosystems (*e.g.* Pauly *et. al.* 2000), relied on trophic-level estimates obtained through Ecopath applications, that is, diet composition studies that were rendered mutually compatible in an ecosystem context. They document the utility of the post-Rigler trophic-level concept.

www.ingramcontent.com/pod-product-compliance
Lightning Source LLC
Chambersburg PA
CBHW061324190326
41458CB00011B/3892